Lecture Notes in Physics

The Editorial Policy for Proceedings

The series Lecture Notes in Physics reports new developments in physical research and teaching – quickly, informally, and at a high level. The proceedings to be considered for publication in this series should be limited to only a few areas of research, and these should be closely related to each other. The contributions should be of a high standard and should avoid lengthy redraftings of papers already published or about to be published elsewhere. As a whole, the proceedings should aim for a balanced presentation of the theme of the conference including a description of the techniques used and enough motivation for a broad readership. It should not be assumed that the published proceedings must reflect the conference in its entirety. (A listing or abstracts of papers presented at the meeting but not included in the proceedings could be added as an appendix.)
When applying for publication in the series Lecture Notes in Physics the volume's editor(s) should submit sufficient material to enable the series editors and their referees to make a fairly accurate evaluation (e.g. a complete list of speakers and titles of papers to be presented and abstracts). If, based on this information, the proceedings are (tentatively) accepted, the volume's editor(s), whose name(s) will appear on the title pages, should select the papers suitable for publication and have them refereed (as for a journal) when appropriate. As a rule discussions will not be accepted. The series editors and Springer-Verlag will normally not interfere with the detailed editing except in fairly obvious cases or on technical matters.
Final acceptance is expressed by the series editor in charge, in consultation with Springer-Verlag only after receiving the complete manuscript. It might help to send a copy of the authors' manuscripts in advance to the editor in charge to discuss possible revisions with him. As a general rule, the series editor will confirm his tentative acceptance if the final manuscript corresponds to the original concept discussed, if the quality of the contribution meets the requirements of the series, and if the final size of the manuscript does not greatly exceed the number of pages originally agreed upon.
The manuscript should be forwarded to Springer-Verlag shortly after the meeting. In cases of extreme delay (more than six months after the conference) the series editors will check once more the timeliness of the papers. Therefore, the volume's editor(s) should establish strict deadlines, or collect the articles during the conference and have them revised on the spot. If a delay is unavoidable, one should encourage the authors to update their contributions if appropriate. The editors of proceedings are strongly advised to inform contributors about these points at an early stage.
The final manuscript should contain a table of contents and an informative introduction accessible also to readers not particularly familiar with the topic of the conference. The contributions should be in English. The volume's editor(s) should check the contributions for the correct use of language. At Springer-Verlag only the prefaces will be checked by a copy-editor for language and style. Grave linguistic or technical shortcomings may lead to the rejection of contributions by the series editors.
A conference report should not exceed a total of 500 pages. Keeping the size within this bound should be achieved by a stricter selection of articles and not by imposing an upper limit to the length of the individual papers.
Editors receive jointly 30 complimentary copies of their book. They are entitled to purchase further copies of their book at a reduced rate. As a rule no reprints of individual contributions can be supplied. No royalty is paid on Lecture Notes in Physics volumes. Commitment to publish is made by letter of interest rather than by signing a formal contract. Springer-Verlag secures the copyright for each volume.

The Production Process

The books are hardbound, and quality paper appropriate to the needs of the authors is used. Publication time is about ten weeks. More than twenty years of experience guarantee authors the best possible service. To reach the goal of rapid publication at a low price the technique of photographic reproduction from a camera-ready manuscript was chosen. This process shifts the main responsibility for the technical quality considerably from the publisher to the authors. We therefore urge all authors and editors of proceedings to observe very carefully the essentials for the preparation of camera-ready manuscripts, which we will supply on request. This applies especially to the quality of figures and halftones submitted for publication. In addition, it might be useful to look at some of the volumes already published.
As a special service, we offer free of charge LATEX and TEX macro packages to format the text according to Springer-Verlag's quality requirements. We strongly recommend that you make use of this offer, since the result will be a book of considerably improved technical quality.
To avoid mistakes and time-consuming correspondence during the production period the conference editors should request special instructions from the publisher well before the beginning of the conference. Manuscripts not meeting the technical standard of the series will have to be returned for improvement.

For further information please contact Springer-Verlag, Physics Editorial Department V, Tiergartenstrasse 17, W-6900 Heidelberg, FRG

T. M. M. Verheggen (Ed.)

Numerical Methods for the Simulation of Multi-Phase and Complex Flow

Proceedings of a Workshop
Held at Koninklijke/Shell-Laboratorium, Amsterdam
Amsterdam, The Netherlands, 30 May - 1 June 1990

Springer-Verlag
Berlin Heidelberg New York
London Paris Tokyo
Hong Kong Barcelona
Budapest

Editor

T. M. M. Verheggen
Koninklijke/Shell-Laboratorium, Amsterdam
Badhuisweg 3
NL-1031 CM Amsterdam
The Netherlands

ISBN 3-540-55278-2 Springer-Verlag Berlin Heidelberg New York
ISBN 0-387-55278-2 Springer-Verlag New York Berlin Heidelberg

Printed in Germany

Typesetting: Camera ready by author
Printing and binding: Druckhaus Beltz, Hemsbach/Bergstr.
58/3140-543210 - Printed on acid-free paper

CONTENTS

Introduction.

T. Verheggen, P. Rem and J. Somers

Koninklijke/Shell-Laboratorium, Amsterdam (Shell Research B.V.)
Postbus 3003, 1003 AA Amsterdam, the Netherlands

From May 30 until June 1 1990 a workshop was held at Koninklijke/Shell-Laboratorium, Amsterdam entitled "Workshop on Numerical Methods for the Simulation of Multi-Phase and Complex Flow". The objective of this workshop was to assess the relative value of lattice gas and lattice Boltzmann models, on the one hand, and 'classical' numerical techniques on the other hand, as tools for the simulation of complex flow phenomena.

The use of lattice gas models for fluid flow simulation dates from 1986 when Frisch, Hasslacher and Pomeau [1] showed that a hexagonal lattice gas was capable of simulating two-dimensional Navier-Stokes flow. Soon afterwards, d'Humières, Lallemand and Frisch [2] showed that a lattice gas model based on the four-dimensional face-centered hypercubic lattice could be used to simulate three-dimensional Navier-Stokes flow. Up to that time, lattice gases had mainly been used as model systems in fundamental studies in statistical mechanics.

Since the introduction of the basic lattice gas models, many extensions have been constructed. One of the most interesting is towards two-phase flow. The first model that conserved the mass of both phases and showed surface tension was invented by Rothman and Keller [3]. In these two-phase models, two different types of particles move on the same lattice. Collisions involving the different types are chosen such that phase separation and surface tension are obtained.

Another interesting line of development is the construction of models for Darcy flow [4]. To achieve Darcy flow the lattice is filled with a small fraction of 'scatterers' from which particles are 'bounced back'.

The Boltzmann lattice gas [5, 6] is a variant of the Boolean lattice gas, in which the state of a cell is defined by the real-valued mean occupancy. The dynamics of the lattice gas is defined by an evolution equation for the mean occupancy incorporating both propagation and collision.

The theory behind all these models is relatively well developed now and has been verified by numerical experiments [7, 8, 9]. Some experience also exists with the use of these models as numerical tools [8, 9, 10, 11].

Though lattice gas models are continuously being developed further, the time now seemed to have come to take stock of where we stand and assess the relative value of 'classical' numerical techniques and lattice gas and lattice Boltzmann models as numerical tools for the simulation of complex flow. We discuss briefly some of the fundamental issues as well as some algorithmic and computational properties of the methods. In addition, we comment on the status of different areas of application. Of course, the discussion will not be comprehensive. It mainly serves to introduce the papers in these proceedings and to refer to some of basic papers in the area.

1 Fundamental Issues.

Simulation techniques based on numerical analysis usually start from macroscopic laws that are either founded in a phenomenological way or are derived from laws at a microscopic level through the theory of statistical mechanics. The macroscopic fields that appear in these laws are then approximated by a finite set of real variables. The laws themselves are translated into evolution rules for these variables. For this approximation many different methods are used: finite difference, finite volume, finite element or spectral methods. The fundamental question is that of mathematical convergence: the results of the computational scheme have to remain bounded and the property of interest of the solution has to converge to the value imposed by the original continuous laws if the number of variables increases. In the case of complex chaotic solutions, the properties of interest will often be of a statistical nature. Nevertheless, numerical schemes are often judged on their convergence properties for deterministic cases. It may be argued that such a criterion is either too strict, because randomizing effects may destroy the effect of some inconsistencies of a numerical scheme, or too loose, as chaotic systems might blow up small non-physical errors to a macroscopic scale. So far, the relation between deterministic convergence properties and statistical convergence of chaotic solutions is poorly understood.

In contrast to the approach of numerical analysis, the lattice gas methods considered here start with physics at the microscopic level. At this level, the physical system of interest is simplified to a system in which particles are constrained to hop synchronously from one node of the lattice to another. In addition, the interactions between particles are simplified to the point where they reflect only the few conservation laws that characterize the original physical problem. Relatively little attention has been paid to the mathematical convergence properties of lattice gas methods. A complication is the noise that is inherent in lattice gases, resulting in a decrease of the order. Because of their simplicity, lattice gases allow thorough statistical analysis. In many cases, the equilibrium distribution as well as transport coefficients can be found explicitly (see in these proceedings the paper by Ernst, Naitoh, Brito).

Most of the theoretical results rest on general assumptions that are familiar from the kinetic theory of gases (see e.g. [12]). We mention here the assumption of semi-detailed balance, the role of invariants in the equilibrium distribution and the assumption of scale separation. Frisch et al. [13] have shown that semi-detailed balance guarantees convergence to the Fermi-Dirac distribution and Dubrulle et al. [14] have shown that, without semi-detailed balance, large discrepancies with the Boltzmann approximation can occur. In his introductory presentation at the workshop, Frisch discussed the relation between semi-detailed balance and the positivity of viscosity. (In the meantime, Henon has shown that models without semi-detailed balance can have negative viscosity.) Due to their simplicity, lattice gases may have several global invariants besides the usual local invariants (e.g. mass, momentum, energy, etc.). These invariants were first discovered by Zanetti [15]. (See also d'Humières et al. [16]). In principle, their presence should be noticeable from the equilibrium distribution. However, effects of this kind turn out to be extremely small in practical simulations. Finally, the assumption of scale separation in these models has a parallel in numerical analysis in that the phenomena of interest in both kinds of model are visible only on a certain scale in space and time with respect to the units of space and time in the model.

The range of physical phenomena that can be described by lattice gas methods is limited by the possibility of finding a suitable discrete representation that obeys the correct physical laws in the macroscopic limit. Even if such a representation has been found, it may be that the physical parameters are limited to a small window in parameter space. Sometimes, however, such limitations reflect the same limitations of resolution that also exist in numerical analysis for the same set of physical phenomena. An example is the resolution required for the simulation of 'high' Reynolds number flow. In these latter techniques, the impossibility of simulating some part of parameter space with a given finite mesh results in instability or inaccuracy. The inherently stable lattice gas methods give up at an earlier stage.

The lattice Boltzmann method is related both to lattice gases and to numerical analysis. It was originally proposed by McNamara and Zanetti [6] and independently by Higuera and Jiménez [5] as an alternative simulation tool for the discrete lattice gases. The method is derived from a discrete lattice gas by assuming the Boltzmann approximation and linearization around the equilibrium solution. It shares many microscopic concepts, such as conservation properties and propagation with discrete lattice gases, but it has all the formal characteristics of a finite difference scheme. In these proceedings, Succi discusses the approximation properties of the scheme. Higuera proposes a modification to get rid of the acoustic modes (a property due to compressibility, shared with lattice gases). A recent suggestion by Succi is the possibility of using ghost fields for sub-grid modelling of turbulence.

2 About the Algorithms.

One of the special characteristics of the lattice gas technique is that it associates only a few boolean variables with each grid point. The dynamics is governed by discrete state transition rules, which are computationally cheap and do not involve a single floating point operation. As a consequence, very large grids can be used in the simulation and a high resolution of the boundary can be obtained. However, for the solution of the flow the resolution of a single lattice node is much less than what is achieved by, for instance, finite difference techniques.

In a lattice gas simulation, real values of pressure and velocity are obtained by averaging the discrete states of the nodes locally in space and time. These values are inherently noisy, and the signal/noise ratio is the smallest in the low velocity limit.

Improving the signal/noise ratio of a lattice gas solution is computationally very expensive. One could increase the viscosity of the model and incorporate more lattice points, or one could stagger multiple lattices in an ensemble-like system. The disappointment of either approach is that the signal/noise ratio improves only with the square root of the effort. This is in contrast to schemes that use floating point variables and hence achieve an accuracy in the representation of the variables that is exponential in the number of bits used. In the lattice Boltzmann scheme the boolean variables of the lattice gas are replaced by floating point ensembles averages. Hence, hydrodynamics without noise is modelled and memory usage here is in this respect as efficient as with, for instance, finite difference techniques.

Currently, all state-of-the-art techniques share the property that the Reynolds number of the flow is bounded by the ability of the grid to resolve the smallest relevant hydrodynamic features. The collision operators of the early lattice gas models realized a (dimensionless) viscosity of O(1). In consequence, much memory and computation time was wasted, using finer grids than necessary to resolve the features of the flow. Much effort has been put into tuning lattice gas collision rules and lattice Boltzmann schemes, so that nowadays viscosities close to zero can be realized and it is the required resolution that determines the grid size.

Both the lattice gas and the lattice Boltzmann methods are inherently dynamic methods and use explicit time stepping. Their time steps are relatively small. The incompressible limit is approached only if fluctuations in time are small compared to spatial features and the speed of sound. The time step is related explicitly to the spatial resolution and the viscosity, and is not a free parameter of the scheme as in implicit methods. Although the time step of the lattice gas and the lattice Boltzmann methods can be one or two orders of magnitude smaller than with implicit schemes, the amount of work per time step can also be one or two orders of magnitude smaller. The small time step makes them unsuitable for stationary problems, of course. A detailed comparison of the computational efficiency of the methods is not available yet. The lattice Boltzmann method seems to be competitive with other methods, whereas the lattice gas methods often lose out in efficiency as Navier-Stokes solvers.

Concerning boundary conditions one can argue that, if the basic numerical scheme is very sophisticated, then the implementation of general boundary conditions becomes more cumbersome. Lattice gas and lattice Boltzmann schemes can easily accommodate various boundary conditions on complex geometries (i.e. stick, inflow, outflow, shear) [17]. The implementation of these boundary conditions benefits from the fact that the primary variables of the lattice gas and the lattice Boltzmann scheme represent also the components of the stress tensor and higher derivatives of momentum density at each node locally. Obviously, these are not all independent degrees of freedom, but are slaved to the density and velocity field. The result is a generic and consistent formulation of boundary conditions, independent of the particular application, as is not known for other techniques.

Another important issue with respect to the implementation of a numerical scheme is the robustness and stability of the algorithms. A lattice gas simulation does not involve any floating point operations and is fundamentally finite. Although a lattice gas is in the above sense very robust, it will inherently not develop any features that cannot be resolved by the grid. Consequently, its solution will deviate from the true Navier-Stokes solution in these cases. The lattice Boltzmann scheme does share some of the robustness properties of the lattice gas technique. However, in order to optimize computational efficiency it is common practice to use lattice Boltzmann models at low

viscosities, i.e. one or two orders of magnitude lower than the best lattice gas model can achieve. Then numerical instability can occur, leading to momentum densities exceeding the (0,1) interval. This may finally result in floating point exceptions. It is possible to improve the robustness by monitoring the variables and introducing artificial viscosity.

Many authors have already demonstrated that the lattice gas technique can very well be implemented on a parallel computer. (See, for example, [10, 17, 18]). Its basic computation is extremely simple, involving a localized access to a regular data structure only. Very large instances of lattice gas automata can be simulated successfully on these parallel computers, without problems of numerical stability being encountered. Just as with other explicit schemes the number of time steps that is needed to resolve a time-dependent flow increases proportionally with the size of the grid in one spatial dimension. So very large cellular automaton simulations can only be performed in realistic times if, besides a large memory, the parallel architecture also provides impressive processing power. Fast processing power can best be obtained from a dedicated hardware implementation. Cellular automaton machines with high update rates have already been constructed [19, 20]. These efforts clearly benefit from the fact that the lattice gas models use integer arithmetic only. However, most hardware implementations are still limited in their pre- and post-processing facilities and the implementation of boundary conditions. Future designs may recover flexibility by integrating state-of-the-art general purpose processing elements in the cellular automaton architecture. It is also expected that major results can be achieved by hardware implementations that incorporate the more sophisticated recent lattice gas models for three-dimensional and two-phase flow simulation.

3 Applications.

With respect to applications the workshop focused on complex time-dependent Navier-Stokes flow, multi-phase flow of immiscible fluids and complex Darcy flow. Another area of application that will be briefly discussed is that of colloidal suspensions and granular fluids.

Rivet et al. [10] have simulated flow around a circular disk for Reynolds numbers of around 100 with lattice gases to study the instability of vortex shedding for this geometry. In these proceedings, Somers reports a lattice gas simulation of a flow of similar complexity through a complex geometry. In both cases, direct simulation of the Navier-Stokes equation is limited in Reynolds number by computation time and available memory. These applications show an interesting interaction of flow features at different spatial scales. Nevertheless, obtaining quantitative data other than pressure drop and velocity profile (such as dissipation rates) has proven very difficult. The reason for this is the poor signal/noise ratio.

By Succi et al. [11] the lattice Boltzmann technique has been used on a variety problems ranging from laminar 2- and 3-D flows to 2-D turbulent flow. The conclusion that can be drawn from this study is that, for purely hydrodynamic problems, the lattice Boltzmann technique is computationally much more efficient than lattice gases and can be competitive with other numerical techniques. The idea of trying to add a sub-grid model in the sense that was discussed during the workshop by Friedrich [21], seems interesting, but this is clearly beyond the current state-of-the-art in lattice gases.

The difficulty of simulating immiscible multi-phase fluids stems from the need to resolve the geometry of the freely moving interface between the fluids. Good results have been obtained, for example, by finite volume techniques on moving meshes [22]. The motion of the mesh is coupled to the flow such that it follows the interface. Proper boundary conditions at the interface couple back its motion to the flow. These schemes in general require special attention to ensure isotropic behaviour. Lattice gas models with coloured species have also proved successful in modelling immiscible flows. Here the interface is represented implicitly by the densities of the different phases on the grid. Some aspects of the interface conditions of two-phase automate are well understood. Surface tension and isotropy have been studied extensively [23]. Most authors propose interfacial collision rules that conserve mass of each kind, and total momentum locally, but it is not completely understood yet how these microscopic rules aggregate into macroscopic conditions for moving interfaces. In addition, the finite size effects of the interface have not received much attention yet. Another open question is the effect of noise on, for instance, coagulation phenomena. A current research topic is the development of two-phase immiscible lattice Boltzmann methods. Special attention is paid to those models which could eliminate the finite thickness of the lattice gas interface and its Brownian motion. Applications for which numerical simulations have been done include the spinodal decomposition of immiscible lattice gases [24] and capillary phenomena in a porous structure [25].

Studies of complex Darcy flow can be distinguished according to the scale of interest. One type of study [11, 26, 27] starts from the scale of the pore size of a porous medium and aims at the explanation of the transition from Navier-Stokes flow on the pore scale to Darcy flow on an intermediate scale. Other studies are concerned with the flow phenomena on the large spatial and time scales of complete oil or water reservoirs. On the pore scale, lattice gas models and lattice Boltzmann techniques have proved successful already. Permeabilities have been predicted which seem to be in agreement with experiments, albeit that there is no general agreement yet about some details of the models. Three-dimensional multi-phase flow in porous media is a current research topic. The capabilities of lattice gas and lattice Boltzmann techniques in resolving complicated boundary geometries clearly play a decisive role for this application.

Darcy models are used to study viscous fingering in a porous medium. See [28] and also the contribution by Lutsko et al. in these proceedings. For large-scale Darcy flow (e.g. at the scale of an oil reservoir) the lattice gas and lattice Boltzmann Darcy models seem less suitable. One drawback is that they cannot deal efficiently with large variations in physical parameter space. If permeabilities vary greatly with space, or viscosity differences of more than two orders of magnitude occur, then other techniques give the best results [29]. Furthermore, in current state-of-the-art reservoir models, physical phenomena are modelled that cannot be incorporated consistently in lattice gas or lattice Boltzmann techniques.

Another area of application for the lattice gases and the lattice Boltzmann scheme is the simulation of colloidal suspensions and granular flows [30]. Again, their ability to model complicated (moving) boundary geometries makes simulations of these flows on a mesoscopic scale possible. If we consider the suspension of solid particles in gas, two regimes can be identified. If the solids are small and light, then Stokesian hydrodynamics applies. The solids are subject to Brownian motion due to collisions between the solid particles and the gas. Whether or not the noise in the lattice gas can induce the correct Brownian behaviour of the solid is not yet known. The other regime considers moderately sized particles in the limit that need not be Stokesian. These simulations can best be performed by the lattice Boltzmann technique, as the required signal/noise ratio limits the computational efficiency of a lattice gas simulation.

Lattice gases and lattice Boltzmann schemes are capable of simulating colloidal suspensions on a mesoscopic scale of up to a few hundred solid particles. From these simulations, quantitative values for bulk properties can be obtained. These values can then be used to validate theories or to fit empirical coefficients in constitutive relations occurring in equations for flow on larger scales. In general, other techniques are then used to solve these equations.

4 Summary

We have described the status of the lattice gas and lattice Boltzmann methods as a numerical technique. They are now well established in calculating the flow of a simple Navier-Stokes fluid. Any extension of the model, however, such as the inclusion of temperature, multiple species or chemical reactions, involves a major research effort. Much remains to be done here. The capability of these methods in resolving complex geometries is, however, unsurpassed.

This symposium has not, of course, given the final definition of the niche for the application of the lattice gas and lattice Boltzmann methods as numerical tools. Its goal was, by bringing together experts from the different fields, to bring about a discussion of the relative value of the different techniques.

References

1. Frisch, U., Hasslacher, B. and Pomeau, Y.: *Phys. Rev. Lett.* **56** (1986) 1505.
2. d'Humières, D., Lallemand, P. and Frisch, U.: *Europhys. Lett.* **2** (1986) 291.
3. Rothman, D.H. and Keller, J.M.: *J. Stat. Phys.* **52** (1988) 1119.
4. Balasubramanian, K., Hayot, F. and Saam, W.F.: *Phys. Rev. A* **36** (1987) 2248.
5. Higuera, F.J.: *in Discrete Kinetic Theory, Lattice Gas Dynamics and Foundations of Hydrodynamics,* R. Monaca ed., World Scientific 1989.
6. McNamara, G.R. and Zanetti, G.: *Phys. Rev. Lett.* **61** (1988) 2332.
7. d'Humières, D., Lallemand, P.: *Complex Systems* **1** (1987) 599.
8. Higuera, F.J. and Succi, S.: *Europhys. Lett.* **8** (1989) 517.
9. Succi, S., Santangelo, P. and Benzi, R.: *Phys. Rev. Lett.* **60** (1988) 2738.
10. Rivet, J.P., Hénon, M., Frisch, U. and d'Humières, D.: *Europhys. Lett.* **7** (1988) 231.
11. Succi, S., Benzi, R. and Higuera, F.J.: *in Lattice Gas Methods for PDE's, Theory, Applications and Hardware*, G.D. Doolen ed., Physica D **47** (1991) 219.
12. Wolfram, S.: *J. Stat. Phys.***45** (1986) 471.
13. Frisch, U., d'Humières, D., Hasslacher, B., Lallemand, P., Pomeau, Y. and Rivet, J.-P.: *Complex Systems* **1** (1987) 649.
14. Drubulle, D., Frisch, U., Hénon, M. and Rivet, J.-P.: *J. Stat. Phys.* **59** (1990) 1187.
15. Zanetti, G.: *Phys. Rev. A* **40** (1989) 1539.
16. d'Humières, D., Qian, Y.H. and Lallemand, P.: *in Proceedings of the Workshop on Computational Physics and Cellular Automata,* Ouro Preto, Brazil, August 1989.
17. Somers, J.A. and Rem, P.C. *in Parallel Computing 1988*, Springer Lecture Notes in Computer Science 384, 1989.
18. Boghosian, B., Taylor, W. and Rothman, D.H.: *in Proceedings of Supercomputing '88, 2: Science and Applications*, Martin, J.L. and Lundstrom, S.S. eds., IEEE 1989.
19. Clouqueur, A. and d'Humières, D.: *Complex Systems***1** (1987) 585.
20. Toffoli, T. and Margolus, N.: *in Lattice Gas Methods for PDE's, Theory, Applications and Hardware*, G.D. Doolen ed., Physica D **47** (1991) 263.
21. Friedrich, R., Arnal, M. and Unger, F.: *in Computational Fluid Dynamics for the Petrochemical Process Industry,* Applied Scientific Research 48, 1991
22. Fyfe, D.E., Oran, E.S. and Fritts, M.J.: *J. Comp. Phys.* **76** (1988) 349.
23. Somers, J.A. and Rem, P.C.: *in Lattice Gas Methods for PDE's, Theory, Applications and Hardware*, G.D. Doolen ed., Physica D **47** (1991) 39.
24. Rothman, D.H. and Zaleski, S.: *Journal de Physique* **50** (1989) 2126.
25. Rothman, D.H.: *in Discrete Kinetic Theory, Lattice Gas Dynamics and Foundations of Hydrodynamics*, ed. R. Monaco, World Scientific 1989.
26. Rothman, D.H.: *Geophysics* **53** (1988) 509.
27. Chen, S., Diemer, K., Doolen, G.D., Eggert, K., Fu, C., Gutman, S. and Travis, B.J.: *in Lattice Gas Methods for PDE's, Theory, Applications and Hardware*, G.D. Doolen ed., Physica D **47** (1991) 97.
28. Hayot, F.: *in Lattice Gas Methods for PDE's, Theory, Applications and Hardware*, G.D. Doolen ed., Physica D **47** (1991) 64.
29. Russell, T.F. and Wheeler, M.F.: *in The Mathematics of Reservoir Simulation*, Ewing, R.E. ed., SIAM 1983.

30. Ladd, A.J.C. and Frenkel, D. *in Cellular Automata and Modeling of Complex Physical Systems*, P. Manneville et al eds., Springer Proceedings in Physics **46** (1989) 242.

This article was processed using the LaTeX macro package with ICM style

Green-Kubo Formulas for Staggered Transport Coefficients in CA-Fluids

R. Brito and M.H. Ernst*

Institute for Theoretical Physics
University of Utrecht
P.O. Box 80.006
3508 TA Utrecht
The Netherlands

1 Introduction

The majority of lattice gas cellular automata (LGCA) have, apart from the usual conserved quantities new spurious ones. The physical set is conserved on account of properly chosen collision rules, but the unphysical or spurious set is an artifact of the the discrete structure of space and time in which the cellular automaton is defined [1]. The spurious conservation laws have no physical analog in the continuum case. Their typical form is:

$$H_{\theta} = \sum_{r,c} (-)^{t+\theta \cdot r} \, a(\mathbf{c}) n(\mathbf{c}, \mathbf{r}, t) \quad , \tag{1}$$

where $a(\mathbf{c})$ is a collisional invariant, i.e. its sum over all m-particles participating in an m-tuple collision is constant. The integer $\boldsymbol{\theta} \cdot \mathbf{r}$ being even or odd characterizes the two sublattices where $\boldsymbol{\theta}$ belongs to a small set of reciprocal lattice vectors, determined by the model. The occupation number $n(\mathbf{c}, \mathbf{r}, t) = 1$ if the site $\mathbf{r}$ and the velocity channel $\mathbf{c}$ is occupied at time t, and 0 otherwise.

In CA-fluids out of equilibrium, these spurious invariants give rise to new slow modes, the staggered densities. The evolution equations for these modes have to be added to the Navier-Stokes equations, including the proper nonlinear couplings between them and the physical ones. In general, the existence of these new modes will change the macroscopic behavior of the CA-fluid. Also, they modify the mode coupling theory, because there are contributions to the long time tails due to coupling between pairs of staggered modes [2].

However, at the level of linear excitations, these staggered modes do not couple to the hydrodynamic modes, nor to modes characterized by different sublattice divisions. If for a given sublattice characterization $\boldsymbol{\theta}$ only a *single* slow staggered mode exists, then this mode is purely diffusive. This situation occurs in the FHP-model [3], the 9-bits model [4] and the FCHC-model [4]. Green-Kubo relations for the staggered momentum diffusivities have been derived by Zanetti

* Permanent address: Dpto. Física Aplicada I. Facultad Ciencias Físicas. Universidad Complutense. 28040 Madrid. Spain.

[5] and by Ernst and Dufty [6]. However, if there exist for a given $\boldsymbol{\theta}$ *two or more* staggered invariants, partly scalars (e.g. number densities), partly vectors (e.g. momentum densities), then the staggered modes may become propagating waves with a propagation speed and a damping constant.

In the present paper we apply the time correlation function of Ref. [6] to two models. The first one is the FHP model [3] with three staggered momentum densities which are purely diffusive. The last is the 8-bits model [7] with two purely diffusive modes plus four propagating ones with an anisotropic speed of propagation.

2 FHP Model: Purely Diffusive Modes

This model is defined in a triangular lattice with 6 or 7 allowed particles per node. The 7-bits model includes a rest particle with $\mathbf{c} = 0$. It has three new conserved quantities, defined as [5]:

$$\sum_{r} g_{\theta}(\mathbf{r}, t) = \sum_{r,c} (-)^{t+\theta \cdot r} \, (\widehat{\boldsymbol{\theta}} \cdot \mathbf{c}) \; n(\mathbf{c}, \mathbf{r}, t) \ , \tag{2}$$

for $\boldsymbol{\theta} = (0, 2/\sqrt{3})$, $\boldsymbol{\theta}' = (-1, -1/\sqrt{3})$, $\boldsymbol{\theta}'' = (1, -1/\sqrt{3})$ where the hat denotes unit vectors. These new modes are staggered in space *and* time. As was pointed out in the introduction, these modes are purely diffusive because there is only one conserved density per $\boldsymbol{\theta}$-vector. Then, in the Fourier variable $\mathbf{k}$, they satisfy:

$$\partial_t g_{\theta}(\mathbf{k}, t) = -k^2 \Lambda_{\theta}(\widehat{\mathbf{k}}) \; g_{\theta}(\mathbf{k}, t) \ . \tag{3}$$

The Green-Kubo formulas [6] give the diffusivity $\Lambda_{\theta}(\widehat{\mathbf{k}})$ in terms of the current correlation function:

$$\Lambda_{\theta}(\widehat{\mathbf{k}}) = \lim_{z \to 0} \lim_{k \to 0} \frac{1}{\langle g_{\theta} | g_{\theta} \rangle} \left\{ \sum_{t=0}^{\infty} e^{-zt} \langle J_{\theta} | J_{\theta}(t) \rangle - \tfrac{1}{2} \langle J_{\theta} | J_{\theta} \rangle \right\} \ , \tag{4}$$

where $J_{\theta}(t)$ is the longitudinal current associated with the staggered mode $\boldsymbol{\theta}$:

$$J_{\theta}(\mathbf{k}, t) = \sum_{c} (-)^t c_l c_{\|} n(\mathbf{c}, \mathbf{k} + \pi \boldsymbol{\theta}, t) \ , \tag{5}$$

where $c_{\|} = \widehat{\boldsymbol{\theta}} \cdot \mathbf{c}$ and $c_l = \widehat{\mathbf{k}} \cdot \mathbf{c}$ and $n(\mathbf{c}, \mathbf{k}, t)$ is the Fourier transform of the fluctuation $\delta n(\mathbf{c}, \mathbf{r}, t) = n(\mathbf{c}, \mathbf{r}, t) - \langle n(\mathbf{c}, \mathbf{r}) \rangle$. The inner product between Fourier components $\langle A | B \rangle$ is defined as:

$$\langle A | B \rangle \equiv \frac{1}{V} \langle A(\mathbf{k}) B^*(\mathbf{k}) \rangle \tag{6}$$

with V the number of nodes in the lattice. Furthermore, $\langle g_{\theta} | g_{\theta} \rangle$ is the susceptibility, given by:

$$\langle g_\theta | g_\theta \rangle = \sum_{c,c'} c_{\|} \langle n(\mathbf{c},\mathbf{k}) | n(\mathbf{c}',\mathbf{k}) \rangle c'_{\|} = \frac{1}{2} \sum_c c^2 \kappa(c) \equiv \chi_g \tag{7}$$

and $\kappa(c)$ is defined as the inner product of the fluctuation in the occupation number [8], i.e.

$$\begin{aligned} \kappa(c) &= \langle n(\mathbf{c},\mathbf{k}) | n(\mathbf{c},\mathbf{k}) \rangle = f(c)(1 - f(c)) \\ f(c) &= \langle n(\mathbf{c},\mathbf{k}) \rangle \ . \end{aligned} \tag{8}$$

As FHP models do not support energy conservation, $f(c)$ is independent of c, so $f(c) = f = \rho/b$, where the bit number b is 6 or 7 for the models FHP-I or III respectively. Also $\kappa(c) = \kappa$ and $\chi_g = \kappa \sum_c c_x^2 \equiv \kappa b c_0^2$ where c_0 is the velocity of sound in the FHP models. The subtracted term in (4), containing the factor 1/2, is called the *propagation* diffusivity. It is a consequence of the discreteness of space and time, surviving at the macroscopic level of the transport coefficients.

In order to have a more convenient notation we introduce the $b \times b$-matrix,

$$\gamma_{cc'}(\boldsymbol{\theta}, \alpha) \equiv \lim_{z \to 0} \lim_{k \to 0} \left\{ \widetilde{\Gamma}_{cc'}(\mathbf{k} + \pi\boldsymbol{\theta}, z + \alpha\pi i) - \tfrac{1}{2}\delta_{cc'} \right\} \ , \tag{9}$$

where c, c' label the different velocity states and $\alpha = \{0, 1\}$. The kinetic propagator is defined as

$$\begin{aligned} \widetilde{\Gamma}_{cc'}(\mathbf{k}, z)\kappa(c') &= \sum_{t=0}^{\infty} e^{-zt} \sum_r e^{-k \cdot r} \langle \delta n(\mathbf{c},\mathbf{r},t) \delta n(\mathbf{c}',\mathbf{0},0) \rangle \\ &= \frac{1}{V} \langle \widetilde{n}(\mathbf{c},\mathbf{k},z) n^*(\mathbf{c}',\mathbf{k},0) \rangle = \langle \widetilde{n}(\mathbf{c},\mathbf{k},z) | n(\mathbf{c}',\mathbf{k},0) \rangle \ , \end{aligned} \tag{10}$$

with $\widetilde{n}(\mathbf{c},\mathbf{k},z)$ the Fourier-Laplace transform of $\delta n(\mathbf{c},\mathbf{r},t)$. The effect of the factors $(-)^t$ and $(-)^{\theta \cdot r}$ in (10) is a shift in the arguments $\mathbf{k}$ and z of the kinetic propagator by $\pi\boldsymbol{\theta}$ and $i\pi$ respectively. In terms of the kinetic propagator, the staggered diffusivity $\Lambda_\theta(\widehat{\mathbf{k}})$ is:

$$\Lambda_\theta(\widehat{\mathbf{k}}) = \chi_g^{-1} \sum_{c,c'} c_l c_{\|} \gamma_{cc'}(\boldsymbol{\theta}, 1) \kappa(c') c'_l c'_{\|} \ . \tag{11}$$

The diffusivity $\Lambda_\theta(\widehat{\mathbf{k}})$, as expressed in (11), can be written as $\Lambda_\theta(\widehat{\mathbf{k}}) = \widehat{k}_\alpha \widehat{k}_\beta \xi_{\alpha\beta}(\widehat{\boldsymbol{\theta}})$, where $\xi_{\alpha\beta}(\widehat{\boldsymbol{\theta}})$ is a second rank tensor depending on $\widehat{\boldsymbol{\theta}}$ and $\mathbf{c}$. The general form of $\xi_{\alpha\beta}(\widehat{\boldsymbol{\theta}})$ is

$$\xi_{\alpha\beta}(\widehat{\boldsymbol{\theta}}) = \xi_\perp \widehat{\theta}_{\perp\alpha} \widehat{\theta}_{\perp\beta} + \xi_{\|} \widehat{\theta}_\alpha \widehat{\theta}_\beta \ , \tag{12}$$

where $\widehat{\boldsymbol{\theta}}_\perp$ is a vector perpendicular to $\widehat{\boldsymbol{\theta}}$ and $\xi_\perp$ and $\xi_{\|}$ are the transverse and longitudinal diffusivities, given by

$$\xi_\perp = \frac{1}{bc_0^2} \sum_{c,c'} c_\perp c_\| \gamma_{cc'}(\boldsymbol{\theta}, 1) c'_\perp c'_\|$$

$$\xi_\| = \frac{1}{bc_0^2} \sum_{c,c'} c_\|^2 \gamma_{cc'}(\boldsymbol{\theta}, 1)(c'_\|)^2 \ , \tag{13}$$

where $c_\perp$ is the component of $\mathbf{c}$ parallel to $\widehat{\boldsymbol{\theta}}_\perp$, i.e. $c_\perp = \widehat{\boldsymbol{\theta}}_\perp \cdot \mathbf{c}$.

In order to estimate the transport coefficients we use the mean field or Boltzmann approximation, in which all correlations between colliding particles are neglected. As shown in Appendix A, this approximation yields for the kinetic propagator in matrix notation:

$$\gamma(\boldsymbol{\theta}, \alpha) = -\left[\frac{1}{\Delta(\boldsymbol{\theta}, \alpha) + \Omega} + \frac{1}{2}\right] \ , \tag{14}$$

where the diagonal matrix $\Delta(\boldsymbol{\theta}, \alpha)_{cc'}$ is defined as:

$$\Delta(\boldsymbol{\theta}, \alpha)_{cc'} = [1 - \exp(i\pi\alpha + i\pi\boldsymbol{\theta} \cdot \mathbf{c})]\delta_{cc'} = [1 - (-)^{\alpha + \boldsymbol{\theta} \cdot \mathbf{c}}]\delta_{cc'} \tag{15}$$

and Ω is the linearized Bolzmann collision operator, defined in Appendix A.

Explicit calculations show that the coefficient $\xi_\perp$ is equal to the kinematic viscosity because

$$(\Delta + \Omega) c_\perp c_\| = \Omega c_\perp c_\| = \lambda_\perp c_\perp c_\| \tag{16}$$

and $\lambda_\perp$ is the eigenvalue associated with the kinematic viscosity. The coefficient $\xi_\|$ is more involved, because $c_\|^2$ is not an eigenvector of $(\Delta + \Omega)$. After some algebra, one finds [9] for FHP models without rest particle,

$$\xi_\| = 9\nu \tag{17}$$

and for models with a rest particle (FHP-II and FHP-III)

$$\xi_\| = 441 \frac{\nu\varpi}{49\varpi + 16\nu} \tag{18}$$

with shear viscosity $\eta = \rho\nu$ and bulk viscosity $\zeta = \rho\varpi$.

3 8-bits Model: Propagating and Diffusive Staggered Modes

The 8-bits model [7] consists of the superposition of two interacting HPP LGCA's. There are eight velocity states: four "slow" ones with velocities $(\pm 1, 0)$ and $(0, \pm 1)$ and four "fast" ones: $(\pm 1, \pm 1)$. Apart from the physical conserved quantities (particle number, momentum and energy), this model has six staggered ones. The corresponding densities are [10]:

$$\widetilde{\rho}_{\theta f}(\mathbf{r},t) = (-)^{\theta\cdot r} \sum_{|c|=\sqrt{2}} n(\mathbf{c},\mathbf{r},t) \qquad \theta = (1,1)$$

$$\rho_{\theta s}(\mathbf{r},t) = (-)^{t+\theta\cdot r} \sum_{|c|=1} n(\mathbf{c},\mathbf{r},t) \qquad \theta = (1,1)$$

$$\rho_{\theta f}(\mathbf{r},t) = (-)^{t+\theta\cdot r} \sum_{|c|=\sqrt{2}} n(\mathbf{c},\mathbf{r},t) \qquad \theta = (0,1),(1,0)$$

$$g_{\theta}(\mathbf{r},t) = (-)^{t+\theta\cdot r} \sum_{c} c_{\|} n(\mathbf{c},\mathbf{r},t) \qquad \theta = (0,1),(1,0) \ . \tag{19}$$

The first density $\widetilde{\rho}_{\theta f}$, denoted with a tilde, is a *geometric* slow mode [11]. It is not staggered in time but only in space. The vector $\boldsymbol{\theta} = (1,1)$ is the same as for the staggered number of slow particles $\rho_{\theta s}$. However, the terms in the Euler or non-dissipative part of the macroscopic equation (see below), which couple $\rho_{\theta s}$ and $\widetilde{\rho}_{\theta f}$, are vanishing because $\rho_{\theta s}$ involves slow particles and $\widetilde{\rho}_{\theta f}$ fast ones. So, both of them are purely diffusive. Associated with $\boldsymbol{\theta} = (1,0)$ and $\boldsymbol{\theta} = (0,1)$ we have two slow modes, $\rho_{\theta f}$ and g_θ, which are both staggered in space and time. Here $\rho_{\theta f}$ is the staggered number of fast particles and g_θ is the staggered momentum of all particles. As we will see later, these modes are propagating.

In the 9-bits model [4], the presence of the rest particle destroys the invariants $\widetilde{\rho}_{\theta f}, \rho_{\theta s}$ and $\rho_{\theta f}$. Only the mode g_θ survives, then being purely diffusive [12].

3.1 Diffusive Staggered Modes

Let us start with the mode $\rho_{\theta s}$. Once we have identified it as purely diffusive, it satisfies a diffusion equation of the form (3) with $\Lambda_\theta(\widehat{\mathbf{k}})$ replaced by a staggered number diffusivity $D_\theta(\widehat{\mathbf{k}})$. The corresponding Green-Kubo formula is given by:

$$\begin{aligned} D_\theta(\widehat{\mathbf{k}}) &= \lim_{z\to 0}\lim_{k\to 0} \frac{1}{\chi_s} \left\{ \sum_{t=0}^{\infty} e^{-zt} \langle J_{\theta s} | J_{\theta s}(t)\rangle - \tfrac{1}{2} \langle J_{\theta s} | J_{\theta s}\rangle \right\} \\ &= \frac{1}{\chi_s} \sum_{c,c'} j_s(c) \gamma_{cc'}(\boldsymbol{\theta},1) \kappa(c') j_s(c') \ . \end{aligned} \tag{20}$$

where the susceptibility χ_s is defined as

$$\chi_s = \langle \rho_{\theta s} | \rho_{\theta s}\rangle = \langle \rho_s | \rho_s \rangle = \sum_{|c|=1} \kappa(c) \ , \tag{21}$$

and the staggered current of slow particles is

$$J_{\theta s}(\mathbf{k},t) = (-)^t \sum_{c} j_s(c) n(\mathbf{c}, \mathbf{k}+\pi\boldsymbol{\theta}, t) \tag{22}$$

with $j_s(c) = c_l\, \delta_{c1}$.

On account of the symmetry of the lattice, the anisotropic diffusivity $D_\theta(\widehat{\mathbf{k}})$ can be splitted into a longitudinal part $D_{\|}$ and a transverse part $D_{\perp}$,

$$D_\theta(\widehat{\mathbf{k}}) = (\widehat{\mathbf{k}} \cdot \widehat{\boldsymbol{\theta}})^2 D_{\|} + (\widehat{\mathbf{k}} \cdot \widehat{\boldsymbol{\theta}}_{\perp})^2 D_{\perp} \tag{23}$$

with

$$D_{\|} = \frac{1}{\chi_s} \sum_{c,c'} c_{\|} \delta_{c1} \gamma_{cc'}(\boldsymbol{\theta}, 1) \kappa(c') c'_{\|} \delta_{c'1} \tag{24}$$

and a similar expression for $D_{\perp}$ with $c_{\|}$ replaced by $c_{\perp}$.

The last purely diffusive mode in the 8-bits model is $\widetilde{\rho}_{\theta f}$. It satisfies a diffusion equation (3) with anisotropic diffusivity $\widetilde{D}_\theta(\widehat{\mathbf{k}})$ given by a formula similar to (20) with $J_{\theta s}$ replaced by

$$\widetilde{J}_{\theta f}(\mathbf{k}, t) = \sum_{c} j_f(c) n(\mathbf{c}, \mathbf{k} + \pi\boldsymbol{\theta}, t) \tag{25}$$

with $j_f(c) = c_l\, \delta_{c\sqrt{2}}$ which is only staggered in space. The counterpart of the second equality in (20) is

$$\widetilde{D}_\theta(\widehat{\mathbf{k}}) = \frac{1}{\chi_f} \sum_{c,c'} j_f(c) \gamma_{cc'}(\boldsymbol{\theta}, 0) \kappa(c') j_f(c') \tag{26}$$

with χ_f defined as

$$\chi_f \equiv \langle \widetilde{\rho}_{\theta f} | \widetilde{\rho}_{\theta f} \rangle = \langle \rho_{\theta f} | \rho_{\theta f} \rangle = \sum_{|c|=\sqrt{2}} \kappa(c) \ . \tag{27}$$

Again the symmetries allow us to separate $\widetilde{D}_\theta(\widehat{\mathbf{k}})$ into a longitudinal part $\widetilde{D}_{\|}$ and a transverse part $\widetilde{D}_{\perp}$ as:

$$\widetilde{D}_\theta(\widehat{\mathbf{k}}) = (\widehat{\mathbf{k}} \cdot \widehat{\boldsymbol{\theta}})^2 \widetilde{D}_{\|} + (\widehat{\mathbf{k}} \cdot \widehat{\boldsymbol{\theta}}_{\perp})^2 \widetilde{D}_{\perp} \tag{28}$$

with

$$\widetilde{D}_{\|} = \frac{1}{\chi_f} \sum_{c,c'} c_{\|} \delta_{c\sqrt{2}} \gamma_{cc'}(\boldsymbol{\theta}, 0) \kappa(c') c'_{\|} \delta_{c'\sqrt{2}} \tag{29}$$

and a similar expression for $\widetilde{D}_{\perp}$ with $c_{\|}$ replaced by $c_{\perp}$. In Boltzmann approximation the matrix $\gamma(\boldsymbol{\theta}, \alpha)$ is given by (14).

3.2 Propagating Staggered Modes

For $\boldsymbol{\theta} = (0,1)$ or $(1,0)$ there are two slow staggered densities: $g_\theta(\mathbf{k},t)$ and $\rho_{\theta f}(\mathbf{k},t)$.

Let us consider first the reversible part (Euler part) of the evolution equations for these two staggered densities. Following the method of Ref. [6], we obtain,

$$\begin{aligned}\partial_t \rho_{\theta f} &= -ikg_\theta \frac{\langle g_\theta | J_{\theta f}\rangle}{\langle g_\theta | g_\theta\rangle} = -ikg_\theta(\widehat{\boldsymbol{\theta}}\cdot\widehat{\mathbf{k}})w_0^2 \\ \partial_t g_\theta &= -ik\rho_{\theta f}\frac{\langle \rho_{\theta f}|J_{\theta g}\rangle}{\langle \rho_{\theta f}|\rho_{\theta f}\rangle} = -ik\rho_{\theta f}(\widehat{\boldsymbol{\theta}}\cdot\widehat{\mathbf{k}}) \ . \end{aligned} \tag{30}$$

The currents $J_{\theta g}$ and $J_{\theta f}$ are defined through (22) with $j_s(c)$ replaced by $j_g(c) = c_l c_{\|}$ and $j_f(c) = c_l\,\delta_{c\sqrt{2}}$ respectively and $w_0^2 = \chi_f/\chi_g$ with χ_f and χ_g defined respectively in (27) and (7).

The solution to (30) represents two propagating staggered waves ($\sigma = \pm$),

$$\psi_\theta^\sigma(\mathbf{k},t) = \rho_{\theta f}(\mathbf{k},t) + \sigma w_0 g_\theta(\mathbf{k},t) \tag{31}$$

with a speed of propagation $c_\theta(\widehat{\mathbf{k}}) = |\widehat{\mathbf{k}}\cdot\widehat{\boldsymbol{\theta}}|w_0$ that depends on the direction of propagation $\widehat{\mathbf{k}}$. Having established the proper linear combinations of staggered modes ψ_θ^σ that diagonalizes the Euler equations (30), the complete time dependence of the staggered excitations is given by [6]:

$$\begin{aligned}\psi_\theta^\sigma(\mathbf{k},t) &= \psi_\theta^\sigma(\mathbf{k},0)\exp[-\omega_\theta^\sigma(\mathbf{k})t] \\ \omega_\theta^\sigma(\mathbf{k}) &= i\sigma(\mathbf{k}\cdot\widehat{\boldsymbol{\theta}})w_0 + k^2\Lambda_\theta^\sigma(\widehat{\mathbf{k}}) \ . \end{aligned} \tag{32}$$

After some algebra the Green-Kubo formula for the damping constant is found to be

$$\begin{aligned}\Lambda_\theta^\sigma(\widehat{\mathbf{k}}) &= \lim_{z\to 0}\lim_{k\to 0}\frac{1}{\langle\psi_\theta^\sigma|\psi_\theta^\sigma\rangle}\left\{\sum_{t=0}^{\infty} e^{-zt}\langle \widehat{J}_\theta^\sigma|\widehat{J}_\theta^\sigma(t)\rangle - \tfrac{1}{2}\langle\widehat{J}_\theta^\sigma|\widehat{J}_\theta^\sigma\rangle\right\} \\ &= \frac{1}{2\chi_f}\sum_{c,c'} j_\theta^\sigma(c)\gamma_{cc'}(\boldsymbol{\theta},1)\kappa(c')j_\theta^\sigma(c') \ , \end{aligned} \tag{33}$$

where the staggered subtracted current $\widehat{J}_\theta^\sigma$ is given by an expression similar to (22) with j_s replaced by

$$j_\theta^\sigma(c) = [c_l - \sigma w_0(\widehat{\mathbf{k}}\cdot\widehat{\boldsymbol{\theta}})][\delta_{c\sqrt{2}} + \sigma w_0 c_{\|}] \ . \tag{34}$$

By further exploiting the symmetries the damping constant can be written as

$$\Lambda_\theta^\sigma(\widehat{\mathbf{k}}) = (\widehat{\mathbf{k}}\cdot\widehat{\boldsymbol{\theta}})^2\Lambda_{\|} + (\widehat{\mathbf{k}}\cdot\widehat{\boldsymbol{\theta}}_\perp)^2\Lambda_\perp \ , \tag{35}$$

where each Λ is the sum of an even part Λ_e and an odd part Λ_o, with

$$
\begin{aligned}
\Lambda_{\perp e} &= \frac{1}{2\chi_g} \sum_{c,c'} c_\perp c_\parallel \gamma_{cc'}(\boldsymbol{\theta},1)\kappa(c')c'_\perp c'_\parallel \\
\Lambda_{\perp o} &= \frac{1}{2\chi_f} \sum_{c,c'} c_\perp \delta_{c\sqrt{2}} \gamma_{cc'}(\boldsymbol{\theta},1)\kappa(c')c'_\perp \delta_{c'\sqrt{2}} \\
\Lambda_{\parallel e} &= \frac{1}{2\chi_g} \sum_{c,c'} (c_\parallel^2 - \delta_{c\sqrt{2}}) \gamma_{cc'}(\boldsymbol{\theta},1)\kappa(c')(c_\parallel'^2 - \delta_{c'\sqrt{2}}) \\
\Lambda_{\parallel o} &= \frac{1}{2\chi_f} \sum_{c,c'} c_\parallel (\delta_{c\sqrt{2}} - w_0^2) \gamma_{cc'}(\boldsymbol{\theta},1)\kappa(c')c'_\parallel (\delta_{c'\sqrt{2}} - w_0^2) \ . \qquad (36)
\end{aligned}
$$

These relations are the Green-Kubo formulas for the damping constant of the staggered propagating waves in models supporting for a given sublattice division both staggered number as well as momentum invariants. In Boltzmann approximation the matrix $\gamma(\boldsymbol{\theta},1)$ is given by (14). In fact, the staggered transport coefficients in the 8-bits model are simply related to usual viscosities and heat conductivity, as will be shown in [10].

Appendix A: Boltzmann Approximation

The time evolution of the occupation numbers $n(\mathbf{c},\mathbf{r},t)$ is given by:

$$
n(\mathbf{c},\mathbf{r}+\mathbf{c},t+1) = n(\mathbf{c},\mathbf{r},t) + I(\mathbf{c}|n) \ , \qquad (37)
$$

where the first term of the left hand side of this equation is the *free streaming* and the term $I(\mathbf{c}|n)$ represents the *collision operator*. In general, $I(\mathbf{c}|n)$ is nonlinear in the occupation numbers, *i.e.*, in a b-bits model, $I(\mathbf{c}|n)$ contains at most b n's, each referring to a different velocity channel. Iterating (37) t times, one obtains the exact series for $n(\mathbf{c},\mathbf{r},t)$ expressed as a polynomial of degree b^t in the occupation numbers $n(\mathbf{c},\mathbf{r}',0)$. In the approximation of uncorrelated collisions, the so-called Boltzmann approximation, in which recollisions between particles are neglected, one can resum the series as a t-th power of a single step evolution.

If we now make the replacement $n(\mathbf{c},\mathbf{r},t) = \delta n(\mathbf{c},\mathbf{r},t) + f(c)$ with $f(c) = \langle n(\mathbf{c},\mathbf{r})\rangle$, the collision term becomes (taking into account that $I(f)$ is equal to 0):

$$
I(\mathbf{c}|n) = \sum_{c'} \Omega_{cc'} \delta n(\mathbf{c}',\mathbf{r},t) + \sum_{c'c''} \Omega_{cc'c''} \delta n(\mathbf{c}',\mathbf{r},t) \delta n(\mathbf{c}'',\mathbf{r},t) + \cdots \qquad (38)
$$

Then, at linear order, the time evolution for uncorrelated collisions becomes:

$$
\delta n(\mathbf{c},\mathbf{r}+\mathbf{c},t+1) = \delta n(\mathbf{c},\mathbf{r},t) + \sum_{c'} \Omega_{cc'} \delta n(\mathbf{c}',\mathbf{r},t) \ . \qquad (39)
$$

After Fourier-Laplace transformation we obtain the solution

$$\widetilde{n}(\mathbf{c},\mathbf{k},z) = \sum_{c'} e^{z+ik\cdot c} \left(\frac{1}{e^{z+ik\cdot c} - 1 - \Omega} \right)_{cc'} n(\mathbf{c}',\mathbf{k},0) \ , \tag{40}$$

where $n(\mathbf{c},\mathbf{k},t)$ and $\widetilde{n}(\mathbf{c},\mathbf{k},z)$ have been defined in section 2 as the Fourier and Fourier-Laplace transform of $\delta n(\mathbf{c},\mathbf{r},t)$.

ACKNOWLEDGEMENTS

One of us (RB) acknowledges the support of the Offices of International Relations of The State University of Utrecht and Universidad Complutense of Madrid and of DGICYT (grant number PB88-0140).

References

1. L.P. Kadanoff, G.R. McNamara and G. Zanetti, *Phys. Rev.* A **40**, 4527 (1989).
2. T. Naitoh, M.H. Ernst and J.W. Dufty, *Phys. Rev.* A (1990).
3. U. Frisch, B. Hasslacher and Y. Pomeau, *Phys. Rev. Lett.* **56**, 1505 (1986).
4. D. d'Humières, P. Lallemand and U. Frisch, *Europhys. Lett.* **2**, 291 (1986).
5. G. Zanetti, *Phys. Rev.* A **40**, 1539 (1989).
6. M.H. Ernst and J.W. Dufty, *J. Stat. Phys* **58**, 57 (1990).
7. B. Chopard and M. Droz, *Phys. Lett.* A **126**, 476 (1988).
8. M.H. Ernst, in *Fundamental Problems in Statistical Mechanics* VII, H. van Beijeren, Ed., (North Holland Publ. Co, Amsterdam, 1990).
9. R. Brito, M.H. Ernst and T.R. Kirkpatrick, *J. Stat. Phys.* (1990).
10. R. Brito, M.H. Ernst, *J. Phys. A*, to be published.
11. D. d'Humières, Y.H. Qian and P. Lallemand, *Discrete Kinetic Theory, Lattice Gas Dynamics and Foundations of Hydrodynamics*, R. Monaco, Ed., (World Scientific, Singapore, 1989).
12. S.P. Das and M.H. Ernst, to appear. M.H. Ernst, S. P. Das and D. Kamman, this volume.

Heat conductivity in a thermal LGCA

M. H. Ernst, S. P. Das and D. Kammann

Instituut voor theoretische fysica, Rijksuniversiteit Utrecht
3508 TA Utrecht, The Netherlands.

1 Introduction

A Lattice Gas Cellular Automaton (LGCA) consists of many interacting particles moving on a regular space lattice with a small set of different speeds. These models have been used as practical approximations for simulating physical systems and are very efficient for computational purposes. Numerical simulations of two and three-dimensional incompressible fluid flows and wave motion have produced good results. However, LGCA's were also used as bona fide statistical mechanics models with extremely simplified dynamics.

In the earlier versions of LGCA's [1] all the particles had the same speed apart from rest particles. Energy was either not conserved or just corresponded to number conservation. Thus *temperature* did not exist for such systems. Later, multiple-speed models were introduced to have microscopic energy conservation and hence temperature in the system. In the present work we consider the 8- or 9-bits LGCA, introduced by Chopard et al [2], and d'Humières [3] et al respectively. The particles move on a square lattice with speeds $|\vec{c}|^2 = 0,1,2$ and will be referred to as rest, slow and fast particles. The collision rules gives rise to non trivial energy conservation. The concept of temperature in the non-equilibrium fluid is introduced through the *local equilibrium* ensemble in close analogy with the theory of continuous fluids. Here the emphasis is on *linear* transport, in particular on the heat current. We evaluate the Green-Kubo expression [4] for the thermal conductivity in the Boltzmann approximation. Thus our consideration here is restricted to effects coming from uncorrelated collisions between the particles. We ignore correlated collisions which gives rise to the so called ring collisions that lead to mode-coupling theories [5].

We only present some details of the calculation of the thermal conductivity of the LGCA, implementing the Boltzmann approximation on the Green-Kubo expressions for the transport coefficients. We state here the analytic results for the two specific models mentioned above. We also present the results for the speed of sound waves in a LGCA for different densities and temperatures. We evaluate numerically for the 9-bits model the expression for the heat conductivity obtained from the Green-Kubo formulas.

2 Description of the model

The model is defined on a two dimensional square lattice. The state of the system is given by the set of occupation numbers $n(\vec{c},\vec{r},t)$. Then $n(\vec{c},\vec{r},t)$ is equal to 1 if the velocity channel $\vec{c}$ at site $\vec{r}$ is occupied and equals 0 otherwise. The collision rules for the multi-speed model that conserve particle number, momentum and energy are shown in figure 1.

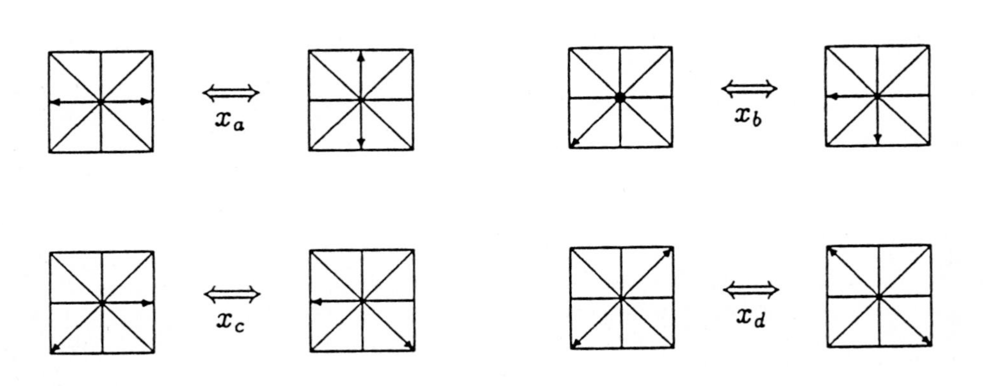

Fig. 1. The Collision rules for the *thermal* LGCA. The disk in (b)-collision represents a rest particle

The probability for the different collision events are denoted by x_a x_b x_c x_d respectively. The 9-bits model corresponds to setting x_a=x_b=x_d=1 and x_c=0, with only binary collisions allowed at each site. The 8-bits model is given by x_a=x_c=x_d=1 and x_b=0, where the a, d collisions can occur simultaneously at the same site.

3 Thermodynamics

In equilibrium, the LGCA is very similar to an ideal Fermi gas. The distribution function or the average occupation of a single particle state with velocity $\vec{c}$ at site $\vec{r}$ is given by

$$< n(\vec{c},\vec{r}) > \doteq f^0(c) = \left[1 + z^{-1}\theta^{-c^2}\right]^{-1} . \tag{1}$$

In the 9-bits model $c^2 = \{0, 1, 2\}$; $f^0(c) = \{f^0(0), f^0(1), f^0(\sqrt{2})\}$ is the Fermi distribution. For the 8-bits model we have $c^2 = \{1, 2\}$; $f^0(c) = \{f^0(1), f^0(\sqrt{2})\}$. $z = exp(\mu/k_B T)$ is the fugacity with μ the chemical potential. The symbol θ is related to the temperature T by $\theta = exp[-(2k_B T)^{-1}]$, where $0 \leq \theta \leq 1$. The density ρ is given by

$$\rho = \sum_c f^0(c) = f^0(0) + 4f^0(1) + 4f^0(\sqrt{2}), \quad 9-\text{bits}(0 \leq \rho \leq 9),$$
$$= 4f^0(1) + 4f^0(\sqrt{2}), \quad 8-\text{bits } (0 \leq \rho \leq 8) \; . \tag{2}$$

The equilibrium correlation of the fluctuations $\delta n(\vec{c}, \vec{r})$ are entirely determined by Fermi statistics and there are no spatial correlations in a lattice gas, *i.e.*

$$< \delta n(\vec{c}, \vec{r}) \delta n(\vec{c}\,', \vec{r}\,') >= \kappa(c) \delta_{cc'} \delta_{rr'}, \tag{3}$$

where $\kappa(c) = f^0(c)(1 - f^0(c))$. In the 9-bits model we have the set $\kappa(c) = \{\kappa(0), \kappa(1), \kappa(\sqrt{2})\}$ and for the 8-bits model $\kappa(c) = \{\kappa(1), \kappa(\sqrt{2})\}$.

4 Sound Speed

Following the usual procedure for a continuous liquid [6] one can obtain an expression for the speed of sound in the LGCA. This involves finding the normal modes of the fluid dynamical equations keeping terms to linear order in the wave vector, the so called Euler terms. For *thermal* lattice gases in d-dimensions the speed of sound c_o is generally given by

$$c_o^2 = \frac{d}{4} \cdot \frac{\sum_c c^4 \kappa(c)}{\sum_c c^2 \kappa(c)}. \tag{4}$$

For the 8-bits or the 9-bits model this reduces to

$$c_o^2 = \frac{1}{2} \cdot \frac{\kappa(1) + 4\kappa(\sqrt{2})}{\kappa(1) + 2\kappa(\sqrt{2})} \; . \tag{5}$$

The quantity on the right hand side is evaluated numerically as a function of the density ρ and the temperature θ and the results are shown in Figure 2.

5 Heat Conductivity

5.1 Green-Kubo formula

The transport coefficients for the LGCA fluid are expressed [7] in the positive diagonal elements of the transport matrix **L**

$$L_{\alpha\beta} = \lim_{s \to 0} V^{-1} \sum_{t=0}^{\infty}{}^{*} e^{-st} < J_\alpha(t) J_\beta >, \tag{6}$$

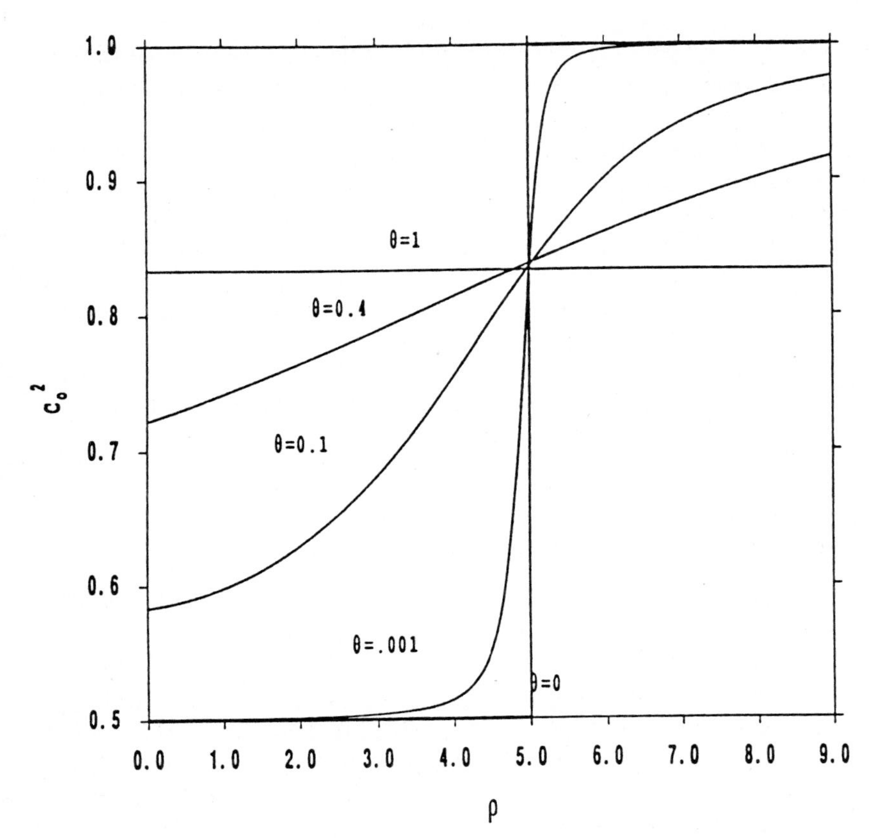

Fig. 2. The speed of sound as a function of density in the 9-bits LGCA fluid for different temperatures

where V is the total number of sites in the system. The * on the summation indicates that the term with $t = 0$ has a weight $\frac{1}{2}$. Green-Kubo relations always contain *subtracted* currents J_α of the general form

$$J_\alpha(t) = \sum_{r,c} j_\alpha(\vec{c})\delta n(\vec{c}, \vec{r}, t), \tag{7}$$

that satisfy the *orthogonality* relations $< J_\alpha A >=0$ for any conserved quantity $A = \{N, \vec{P}, H\}$ in *thermal* (energy conserving) models or $A = \{N, \vec{P}\}$ in *athermal* (energy non-conserving) models. N is the total number of particles, $\vec{P}$ is the total momentum, and H is the energy of the system. Here $\delta n(\vec{c}, \vec{r}, t)$ is the fluctuation of the occupation number $n(\vec{c}, \vec{r}, t)$ around total equilibrium.

The heat conductivity is defined as the coefficient of proportionality in the constitutive relation between the dissipative heat current and the temperature gradient, *i.e.* $\vec{q} = -\lambda\vec{\nabla}T \equiv -2L_{TT}\vec{\nabla}\ln\theta$ with $L_{TT} = k_B T^2\lambda$. In all thermal d-dimensional LGCA's with point particles the conductivity is given by L_{TT} in equation (6) with the subtracted heat current J_T in (7) defined through

$$j_T(\vec{c}) = c_x(\frac{1}{2}c^2 - h), \tag{8}$$

and satisfying $< J_T H >= 0$. The function $h = (2/d)c_o^2$ is similar to the enthalpy per particle, appearing in the corresponding Green-Kubo formula for continuous fluids.

Before focusing on the heat conductivity we point out an interesting difference in the *bulk viscosity* between *athermal* and *thermal* LGCA's. In *multi-speed* models of the former class the bulk viscosity is in general non-vanishing. However, in all models of the latter class the bulk viscosity vanishes identically. The reason is that the unsubtracted current for the bulk viscosity is essentially the energy. If energy is conserved in the model, the subtracted current vanishes identically.

5.2 Boltzmann approximation

The time evolution of the LGCA can be expressed by the microdynamic equation,

$$n(\vec{c}, \vec{r}+\vec{c}; t+1) = n(\vec{c}, \vec{r}; t) + I_c(n(\vec{r}; t)) \; , \tag{9}$$

where I_c denotes the collision term in the velocity channel $\vec{c}$. It depends in a nonlinear fashion on the occupation numbers $n(\vec{c}, \vec{r}; t)$ at site $\vec{r}$ with $\vec{c} = \{\vec{c}_1, \vec{c}_2, \ldots.., \vec{c}_b\}$, b being the number of velocity states in the LGCA. In the Boltzmann approximation we replace the non-equilibrium average $< I_c(n(\vec{r}; t)) >_{ne}$ by $I_c(f)$ where $f(\vec{c}, \vec{r}; t) = < n(\vec{c}, \vec{r}; t) >_{ne}$. Thus the average of the product of the occupation numbers is factorized into a product of averages. This means that the collisions of the particles are treated as being completely random and recollisions are neglected. Taking average of equation (7) thus yields,

$$f(\vec{c}, \vec{r}+\vec{c}; t+1) = f(\vec{c}, \vec{r}; t) + I_c(f) \; . \tag{10}$$

We expand $I_c(f)$ in the deviation from total equilibrium to obtain

$$I_c(f) = I_c(f^0) + \Omega_{cc'}\delta f_{c'} + \Omega_{cc'c''}\delta f_{c'}\delta f_{c''} + \ldots\ldots., \tag{11}$$

where $I_c(f^0) = 0$, and δf_c is short for $\delta f(\vec{c}, \vec{r}, t) = f(\vec{c}, \vec{r}, t) - f^0(c)$. $\Omega_{cc'}$ is the linearized Boltzmann collision operator. From the collision rules defined in section 2, we can construct $I_c(f)$ and hence $\Omega_{cc'}$ explicitly. As shown in Reference [8], the Green-Kubo formula reduces in the Boltzmann approximation to,

$$L_{\alpha\beta} = -\sum_{c',c} j_\alpha(\vec{c}\,') \left[\frac{1}{\Omega} + \frac{1}{2}\right]_{c'c} \kappa(c) j_\beta(\vec{c}) \; . \tag{12}$$

Let $i = 1, 2, 3, 4$ label velocity states with $|\vec{c}_i| = 1$ and let $F_i = f_i/(1 - f_i)$ where $f_i = f(\vec{c}_i, \vec{r}, t)$. It is then straight forward to construct the collision terms for the collisions defined in figure 1. For instance the (a)-collision term for the speed-one particles is given by

$$I_i^{(a)}(f) = x_a \left[F_{i+1}F_{i-1} - F_i F_{i+2}\right] \prod_{j=1}^{b} [1 - f_j] \; . \tag{13}$$

The collision operator is the sum of the relevant diagrams in figure 1. Thus we have

$$I_c(f) = \sum_s I_c^{(s)}(f),$$
$$\Omega = \sum_s \nu_s \Omega^{(s)}, \tag{14}$$

where $s = \{a, b, c, d\}$ labels the diagrams in figure 1. For the 9-bits model the different collision frequencies ν's are defined as

$$\nu_a = \nu_o x_a, \quad \nu_b = \nu_o x_b, \quad \nu_c = \nu_o x_c \theta, \quad \nu_d = \nu_o x_d \theta^2$$
$$\text{and} \quad \nu_o = z^2\theta^2 \prod_{c=1}^{b} [1 - f^0(c)]. \tag{15}$$

We define the eigenfunctions $u_l(c)$, which are b-vectors, and the eigenvalues $\lambda_l^{(s)}$ of the $b \times b$-matrix $\Omega^{(s)}$ through

$$\Omega^{(s)} \kappa u_l = -\lambda_l^{(s)} u_l, \tag{16}$$

where $\kappa_{cc'} = \kappa(c)\delta_{cc'}$ is considered as a diagonal matrix. It follows from equation (14) that the eigenvalues of the collision operator Ω are denoted by $\lambda_l = \sum_s \nu_s \lambda_l^{(s)}$. The norm of u_l is $N_l = \sum_c {u_l}^2(c)$. In table 1 we list these eigenfunctions, the norms and the eigenvalues for the complete set.

	Eigenfunction	Norm	Eigenvalues for collision matrix $\Omega^{(s)}\kappa$			
l	u_l	N_l	λ_l^a	λ_l^b	λ_l^c	λ_l^d
1	1	9	0	0	0	0
2	c_x	6	0	0	0	0
3	c_y	6	0	0	0	0
4	$c^2 - \frac{4}{3}$	4	0	0	0	0
5	$c_x c_y$	4	0	1	4	4
6	$c_x^2 - c_y^2$	4	4	0	0	0
7	$(c^2 - \frac{5}{3})c_x$	$\frac{4}{3}$	0	3	6	0
8	$(c^2 - \frac{5}{3})c_y$	$\frac{4}{3}$	0	3	6	0
9	$(c_x^2 - \frac{2}{3})(c_y^2 - \frac{2}{3})$	$\frac{16}{9}$	0	9	0	0

Table 1. The Eigenfunctions and the Eigenvalues of the operator $\Omega^{(s)}\kappa$.

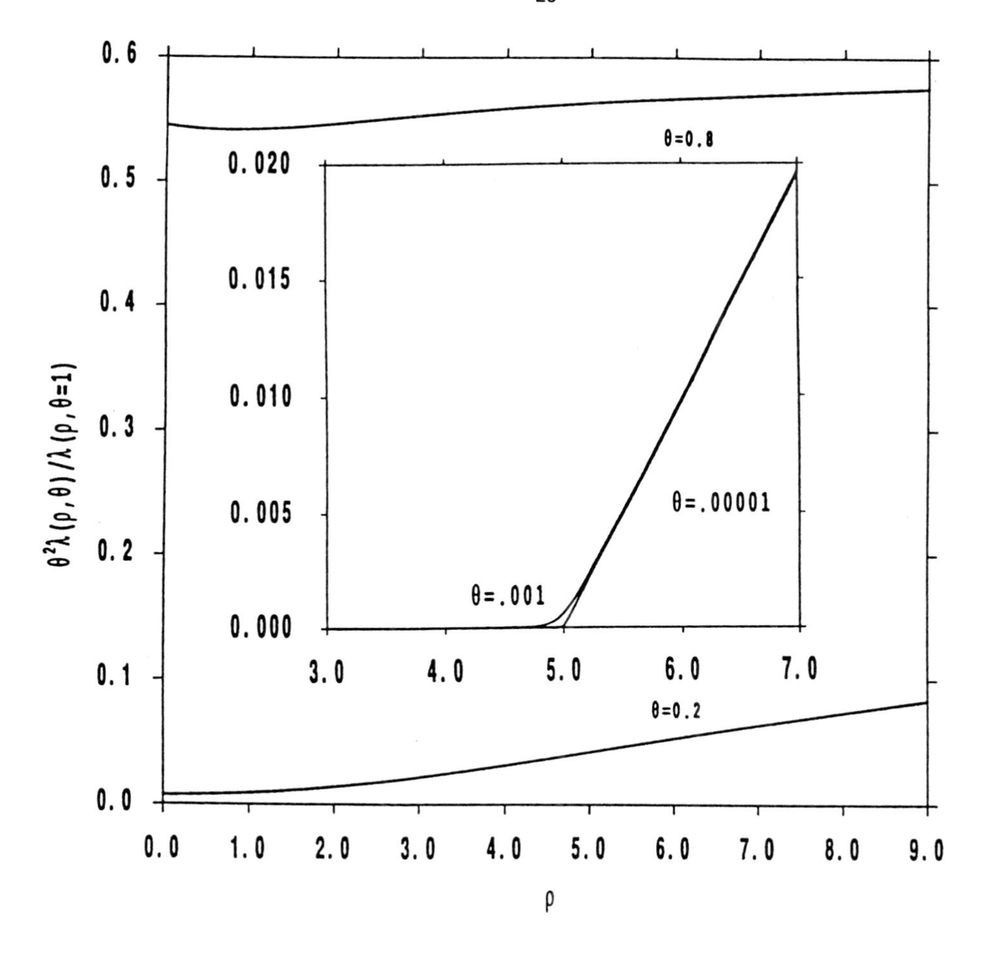

Fig. 3. $\theta^2\lambda(\rho,\theta)$ normalized with respect to its value at $\theta = 1$ vs. the density ρ for the 9-bits model at different temperatures.

Now, we use the orthogonal eigenfunctions u_l's as a basis to expand the quantity $\frac{1}{\Omega}\kappa j_T$ in equation (12),

$$-\frac{1}{\Omega}\kappa j_T = \kappa \sum_m a_m u_m \ . \tag{17}$$

The left hand side of equation (17) is necessarily orthogonal to the zero eigenfunctions $u_1, ..., u_4$ in table 1. Furthermore for $j_T(\vec{c})$ given in equation (8), the symmetry properties of the Ω matrix guarantee that only $u_7(c)$ contributes in the Green-Kubo expression for the thermal conductivity given by equation (12). Using the orthogonality of the eigenfunctions u_l's a_7 can be obtained as

$$a_7 = \frac{< u_7 | j_T >}{\lambda_7 N_7}. \tag{18}$$

Here we have defined the thermal inner product as follows,

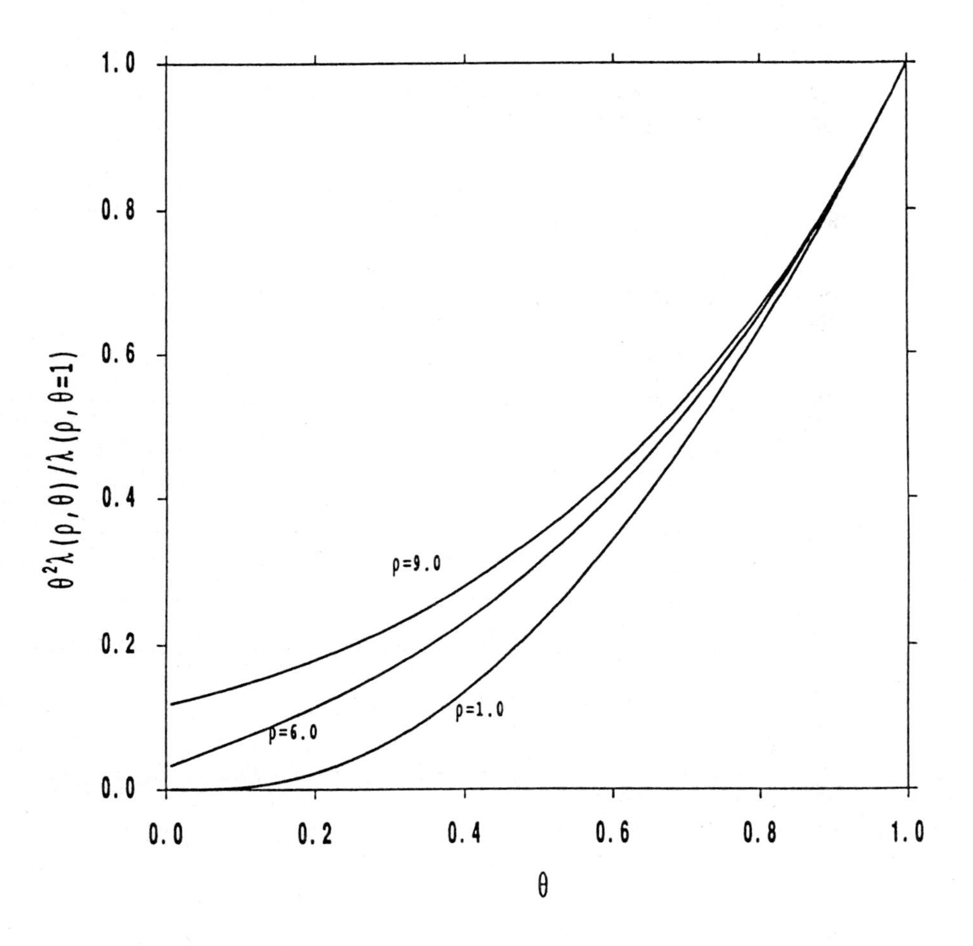

Fig. 4. Temperature dependence of $\theta^2\lambda(\rho,\theta)$ normalized with respect to its value at $\theta = 1$ in the 9-bits model at different densities.

$$< a|b >= \sum_c \kappa(c) a(c) b(c) \; . \tag{19}$$

Now, from equation (12) we obtain the thermal conductivity λ of the LGCA as

$$L_{TT} = k_B T^2 \lambda = < j_T | j_T > \left\{ \frac{3 < j_T | j_T >}{\lambda_7} - \frac{1}{2} \right\} . \tag{20}$$

Explicit calculation with the eigenvectors obtain,

$$< j_T | j_T >= \frac{\kappa(1)\kappa(\sqrt{2})}{\kappa(1) + 2\kappa(\sqrt{2})} \; . \tag{21}$$

The eigenvalue λ_7 is dependent on the model. We give below the results for the 9 and 8 bits models.

$$\begin{aligned} \lambda_7 &= \frac{3z^2\theta^2}{(1+z)(1+z\theta)^4(1+z\theta^2)^4} \, , \qquad (9 - \text{bits model}) \\ &= \frac{6z^2\theta^3}{(1+z\theta)^4(1+z\theta^2)^4} \, , \qquad (8 - \text{bits model}) \; . \end{aligned} \tag{22}$$

We evaluate numerically the expression for the heat conductivity in the 9-bits model. In figure 3 we show how the quantity $\theta^2\lambda(\rho,\theta)$ behaves with density. It is normalized with respect to its value at $\theta = 1$ in order to remove effects coming from the obvious density dependence of the collision frequencies ν_s's. The figure indicates that for moderate to high temperatures, they produce the dominant effect, leaving the normalized function almost structureless with respect to density dependence. However for vanishingly small values of θ, i.e at very low temperature there appear discontinuities in the ρ-dependence of the thermal conductivity around the density ρ=5. In figure 4 we show the isochores for the heat conductivity as a further illustration.

6 Discussion

We have computed the heat conductivity in a *thermal* LGCA fluid in the Boltzmann approximation. The collision laws which involve energy exchange give rise to nontrivial microscopic energy conservation in the system. In fact inspection of table 1 shows that only the energy exchanging collisions (b) and (c) contribute to the eigenvalue λ_7 in equation (20) that determines the heat conductivity.

For two reasons it is more attractive to construct a temperature dependent lattice gas on a triangular lattice, as has been done already by Grosfils and Boon [9]. Firstly, the nonlinear convection term in the Euler equation, involving a fourth rank tensor, does have the isotropic symmetry of the fluid dynamic equations for models with hexagonal symmetry, but is missing this symmetry on the square lattice. Secondly, the fourth rank viscosity tensor in the dissipative part of the Navier-Stokes equation is isotropic on the triangular lattice, but non-isotropic on the square lattice.

Thus in the LGCA models described in this paper, isotropy is not automatic since they are defined on a square lattice. In fact, for the two reasons described above there are two constraints [7] which ensure that the two fourth rank tensors, one in the Euler equation and the other in the dissipative part, are isotropic. The first constraint can only be satisfied in models with multi-speed particles and it restricts the two-dimensional (ρ, θ) phase-space to a line $\theta(\rho)$. The second condition relates the shear viscosity to the extra transport coefficient that appears in the fourth rank viscosity tensor due to the cubic symmetry. Since we can obtain these transport coefficients in the Boltzmann approximation from the Green-Kubo relations given in equation (12), this condition reduces to another relation between ρ and θ, and further restricts the allowed (ρ, θ) parameter space for the system. In fact our numerical calculation shows that it is only for a few points in the (ρ, θ) plane where both these constraints can be satisfied. Detailed results relating to the consideration of isotropy will be reported elsewhere [10].

Acknowledgements

The work of one of us (S.P.D.) is supported by the Stichting voor Fundamenteel onderzoek der Materie (FOM), which is sponsored by NWO.

References

[1] Frish, U., Hasslacher, B. and Pomeau, Y.: Phys. Rev. Lett. **56** (1986) 1505; Frish, U., d'Humières, D., Hasslacher, B., Lallemand, P., Pomeau, Y. and Rivet, J.-P., Complex Syst. **1** (1987)648.

[2] Chopard, B., Droz, M.: Physics Letters A, **126** (1988) 1476.

[3] d'Humières, D., Lallemand, P.: Europhysics Letters, **2** (1986) 291.

[4] Dufty, J.W., Ernst, M. H.: Journal of Stat. Physics, **58** (1990) 57.

[5] Frenkel, D., Ernst, M. H.: Phys. Rev. Lett., **63** (1989)2165.

[6] Hansen, J. P. and McDonnald, I.R.: *Theory of Simple Liquids*, Academic Press, London, 2nd edition 1986.

[7] Ernst, M. H. in *Fundamental Problems in Statistical Mechanics VII*, H. van Beijeren, Ed., (North Holland Publ. Co. Amsterdam, 1990) *Page 321.*

[8] See the article by Brito, R. and Ernst, M. H. in the present volume.

[9] Boon, J. P. and Grosfils, P. private communication.

[10] Ernst, M.H., Das, S. P., to be published.

This article was processed using the LaTeX macro package with ICM style

Effects of Sound Modes on the VACF in Cellular Automaton Fluids

Toyoaki Naitoh[1] , Matthieu H. Ernst[1], Martin A. van der Hoef[2] and Daan Frenkel[2]*

[1] Institute of Theoretical Physics, University of Utrecht, P.O.Box 80006 3508 TA, Utrecht, The Netherlands
[2] FOM Institute for Atomic and Molecular Physics, Kruislaan 413 1098 SJ, Amsterdam, The Netherlands

1 Introduction

The velocity auto correlation function(VACF) of a tagged particle has a fundamental significance in non-equilibrium statistical mechanics. Its time integral defines the self diffusion coefficient through the Green-Kubo relation, and its decay exhibits short time kinetic relaxation, conceivably an intermediate cage effect, and long time hydrodynamic relaxation.

Recently Frenkel, van der Hoef and Ernst [1-4] have obtained the VACF in the 2-D FHP-III model and the quasi 3-D FCHC model with remarkable statistical accuracy using lattice gas cellular automata (LGCA). Their results show excellent agreement with the asymptotic long time tails of $t^{-d/2}$ predicted by mode coupling theory for times larger than $20 \times t_0$ with t_0 being the mean free time. Mode coupling theory assumes that the long time relaxation to equilibrium can be described through the decay of products of hydrodynamic modes [5, 6]. In particular, the combination of a shear and a self diffusion mode leads to the asymptotic long time tail of the VACF [2, 4].

Obviously, the VACF should approach zero for long times, possibly through power law decay. However, when performing Molecular Dynamics (MD-) simulations on a one-dimensional CA-fluid, the VACF of a tagged particle appeared to approach to *a negative constant*, as illustrated in Fig. 1 for a system of 500 lattice sites on a line. It was this remarkable observation, suggesting the existence of some type of conserved quantity, that motivated the present research. Is this a typical one-dimensional effect or could it conceivably also be observed in higher dimensional systems?

Our objective here is not only to give a quantitative explanation of this constant anticorrelation, but also to investigate the VACF in the transition region from kinetic relaxation to long time tails, using mode coupling theory. Following Erpenbeck and Wood [7] the mode coupling theory will be extended (i) to include *all possible product of pairs of hydrodynamic modes* (also those yielding subleading asymptotic time behavior), and (ii) to obtain *finite size corrections* by

---

* On the leave of absence from Senshu University, Higahi-mita, Tama-Ku, Kawasaki, 214, Japan.

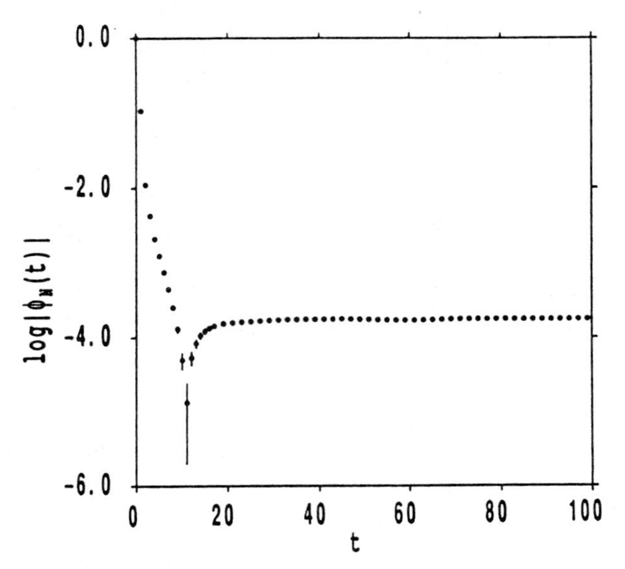

Fig. 1. The simulated VACF in the 1-D self dual 5 bits model at $f = 0.7$ and $L = 500$. The logarithm of $|\phi_N(t)|$ is shown versus time t. Note that the $\phi_N(t)$ is *negative* for $t \geq 12$.

adapting mode coupling theory to finite systems having a discrete set of allowed wave numbers, instead of a continuous range. This extended mode coupling theory will be applied in subsequent sections to 1- and 2-dimensional CA-fluids and compared with existing and new computer simulations. We will give here a rather sketchy analysis of the problem. A more detailed and technical analysis will be published elsewhere [8].

2 Extended Mode Coupling Theory

In computer simulations, the VACF is calculated using the Molecular Dynamics (MD) ensemble, which has a *fixed* number of particles N, one of which is tagged ($N^* = 1$), and *vanishing* total momentum, $\mathbf{P} = 0$. In the time correlation function method and in mode coupling calculations it is most convenient to use a grand equilibrium ensemble, in which all *extensive conserved variables* $\mathbf{A} = \{N, N^*, \mathbf{P}\}$ are fluctuating, but where the average values $\langle N \rangle = \rho V$, $\langle N^* \rangle = 1$ and $\langle \mathbf{P} \rangle = 0$ are prescribed. As we are dealing with lattice gases, V is the total number of lattice sites and ρ is the number of particles per site in the lattice.

In the MD emsemble the VACF is defined as

$$\phi(t) = \frac{\langle v_x(t) v_x(0) \rangle_{MD}}{\langle v_x^2(0) \rangle_{MD}} = \frac{\langle J_x(t) J_x(0) \rangle_{MD}}{\langle J_x^2(0) \rangle_{MD}} \tag{1}$$

with a tagged particle current,

$$J_x(t) \equiv v_x(t) = \sum_{rc} c_x n^*(\mathbf{c}, \mathbf{r}, t)$$

$$= \sum_{rc} c_x n(\mathbf{c}, \mathbf{r}, t) \sigma(\mathbf{c}, \mathbf{r}, t) \quad . \tag{2}$$

In the above equations $n(\mathbf{c}, \mathbf{r}, t)$ is the occupation number of a particle and $n^*(\mathbf{c}, \mathbf{r}, t) = n(\mathbf{c}, \mathbf{r}, t)\sigma(\mathbf{c}, \mathbf{r}, t)$ of a tagged particle. The boolean variable $\sigma(\mathbf{c}, \mathbf{r}, t)$ is 1 for "tagged" and 0 for "untagged". For a b-bits model the reduced density $\langle n \rangle \equiv f = \rho/b$ with $0 \leq f \leq 1$, and $\overline{\sigma} \equiv \langle n^* \rangle / \langle n \rangle = 1/\langle N \rangle$ is the fraction of tagged particles.

According to the time correlation function method the Green-Kubo formula for the diffusion coefficient of a tagged particle is given by the time sum over the VACF, *i.e.*

$$\phi(t) = \left\langle \widehat{J}_x(t) \widehat{J}_x(0) \right\rangle / \left\langle \widehat{J}_x^2(0) \right\rangle \tag{3}$$

defined in the grand ensemble, as indicated above. Here $\hat{J}$ represents a subtracted current [5], *i.e.* a current from which any component parallel to the fluctuations $\delta\mathbf{A} = \mathbf{A} - \langle \mathbf{A} \rangle$ in the conserved quantities has been subtracted,

$$\begin{aligned} \widehat{J}_x(t) &\equiv J_x(t) - \delta\mathbf{A} \cdot \langle \delta\mathbf{A}\delta\mathbf{A} \rangle^{-1} \cdot \langle \delta\mathbf{A} J_x \rangle \\ &= \sum_{\mathbf{rc}} c_x [n^*(\mathbf{c}, \mathbf{r}, t) - \overline{\sigma} n(\mathbf{c}, \mathbf{r}, t)] = v_x(t) - \overline{\sigma} P_x \quad . \end{aligned} \tag{4}$$

In the present case the velocity of each particle has a component $\overline{\sigma}\mathbf{P}$ of order $1/\langle N \rangle$, parallel to the conserved total momentum $\mathbf{P}$. The theory on the ensemble dependence of fluctuations [9, 10], shows that subtracted fluctuation formulas $\left\langle \widehat{F}\widehat{F} \right\rangle$ are ensemble independent in the thermodynamic limit, where $\widehat{F}$ is defined by replacing J in the first line of Eq (4) by F. The normalization in Eq (3) is given by

$$\left\langle \widehat{J}_x^2(0) \right\rangle = V \langle n^* \rangle (1 - \overline{\sigma}) \sum_c c_x^2 \equiv (1 - \overline{\sigma}) c_0^2 \quad . \tag{5}$$

Note that in the thermodynamic limit, where $\overline{\sigma} = 1/\langle N \rangle$ vanishes, the VACF in Eqs (1) and (3) are identical.

In mode coupling studies on the long time behavior of the VACF in Eq (3) the current $\hat{J}$ is interpreted as $\sum_{\mathbf{r}} u_x(\mathbf{r}, t) P(\mathbf{r}, t)$ with $u_x(\mathbf{r}, t)$ the local fluid velocity and $P(\mathbf{r}, t)$ the concentration of tagged particles. The time dependence of these slowly varying fields is calculated from the diffusion equation and the hydrodynamic equations, where the flow field is decomposed into shear modes and sound modes. In our extended mode coupling theory we also include the more rapidly decaying sound modes, which are the only hydrodynamic modes in one-dimensional fluids. This yields [1-4]:

$$\begin{aligned} \phi_N(t) = \frac{(1-f)}{dN} \Bigg\{ & (d-1) \sum_{\mathbf{q} \neq 0} \exp\left[-(D+\nu) q^2 t\right] + \\ & + \sum_{\mathbf{q} \neq 0} \cos(c_0 q t) \exp\left[-(D+\Gamma) q^2 t\right] \Bigg\} \quad , \end{aligned} \tag{6}$$

where D, ν and Γ are the self diffusion coefficient, the kinematic viscosity and the sound damping constant. From here on N stands for the *average* number $\langle N \rangle$ of particles in the system. The dependence on $\overline{\sigma}$, shown in the normalization of Eq (5), cancels completely. In the above equation the summation over $\mathbf{q}$ is restricted to the first Brillouin zone excluding the term of $|\mathbf{q}| = 0$ because $\hat{J}$ is a subtracted current. Hereafter the results obtained through Eq (6) are referred to as the finite hydrodynamics results of the extended mode coupling theory [7]. The finite lattice sum can be calculated numerically and the result is illustrated in Fig. 2 for the one-dimensional model of section 3, with a speed of sound $c_0 = \sqrt{2}$, contained in a volume of $V = L = 500$ lattice sites with perodic boundary conditions. The acoustic traversal time is here $\tau_a = L/c_0 = 354$.

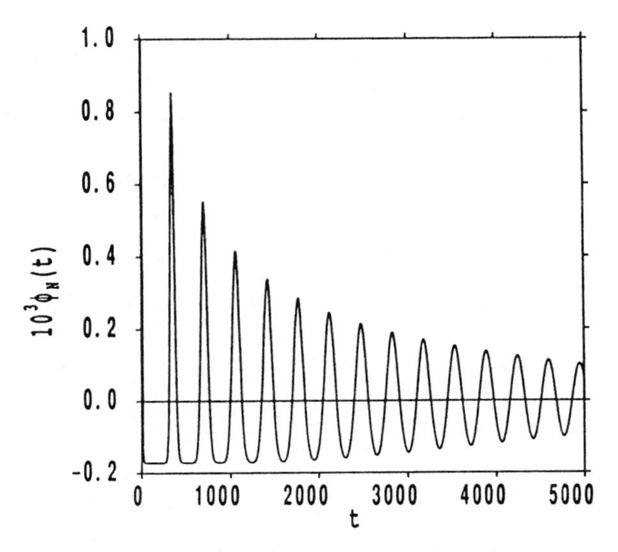

Fig. 2. The finite hydrodynamics prediction for the VACF at $f = 0.7$ and $L = 500$.

In the thermodynamic limit ($V \to \infty$) the $\mathbf{q}$-sum in Eq (6) is replaced by an integral, and the extended mode coupling result for the infinite system becomes [8]:

$$\phi(t) = \frac{(1-f)v_0}{bfd(4\pi t)^{d/2}} \left\{ (d-1)\left(\frac{1}{D+\nu}\right)^{d/2} + \right.$$

$$\left. + \left(\frac{1}{D+\Gamma}\right)^{d/2} {}_1F_1\left(\frac{d}{2}, \frac{1}{2}, -\frac{c_0^2 t}{4(D+\Gamma)}\right) \right\} , \tag{7}$$

where v_0 is the volume of the unit cell and ${}_1F_1$ is the confluent hyper-geometrical function. The first term of Eq. (7) is the leading long time tail obtained in [1] (absent in the case of $d = 1$) and the second one represents the subleading contributions involving the sound modes. For dimensionality $d = 1$, 3 the confluent hypergeometrical function can be expressed in closed form: *i.e.* ${}_1F_1(1/2, 1/2, -z) = e^{-z}$ for $d = 1$ and ${}_1F_1(3/2, 1/2, -z) = (1 - 2z)e^{-z}$ for $d = 3$, and therefore the sound mode contributions are exponentially damped. For dimensionality $d = 2$, ${}_1F_1(1, 1/2, -z)$ can not be represented in any closed form and its asymptotic behavior is estimated as $-1/(2z)$ at large z-values and therefore the decay of the sound mode contribution is proportional to t^{-2}, as compared to the long time contribution, which is proportional to $1/t$. The tranport coefficients are calculated in the Boltzmann approximation.

3 Application to a Fluid on a Line

Our applications start with a 1-D self dual 5 bits model introduced by d'Humières et. al [11]. Only sound modes contribute to Eq (6) and finite size effects are expected to be more significant than in higher dimensional models. The system in this model consists of $V = L$ lattice points on a line. At each lattice point, there are 5 velocity channels, ($b = 5$), each of which can be occupied by at most one particle. The collision rules for these particles are self dual (particle-hole symmetry),

$$
\begin{aligned}
(-1) + (+1) + (0)^\dagger &\Longleftrightarrow (-2) + (+2) + (0)^\dagger \\
(0) + (-1) + (+2)^\dagger &\Longleftrightarrow (+1) + (-2) + (+2)^\dagger \\
(0) + (+1) + (-2)^\dagger &\Longleftrightarrow (-1) + (+2) + (-2)^\dagger \quad ,
\end{aligned}
\tag{8}
$$

where $(i)^\dagger$ represents collisions taking place irrespective of the presence of a "spectator" particle with velocity i at the same lattice point. The collision rules for a tagged particle are only different from those of fluid particles in the sense that the outgoing tagged particle is scattered with equal probability into any allowed outgoing velocity channel.

The simulated VACF in Fig. 1 shows three different types of behavior; an exponential decay for $t \leq 2$ which is well described by the Boltzmann approximation, another exponential decay for $3 \leq t \leq 11$ which is expected to be explained by contributions from sound modes, and a *negative* plateau for $t \geq 20$ which shows finite size effects.

Next we interpret the 1-D simulation data of Fig. 1 and the theoretical result of Fig. 2 which contain only sound mode contributions. It is instructive to transform the finite hydrodynamics result $\phi_N(t)$ back to real space, yielding a representation which is very well suited to discuss the behavior of $\phi_N(t)$ for times shorter than the acoustic traversal time $\tau_a = L/c_0$ [12],*i.e.*

$$
\begin{aligned}
\phi_N(t) = \quad & -\frac{1-f}{N} + \frac{(1-f)}{bf} \times \\
& \times \left\{ \frac{1}{\sqrt{4\pi(D+\Gamma)t}} \sum_{n=-\infty}^{n=\infty} \exp\left[-\frac{L^2}{4(D+\Gamma)}\left(\frac{c_0 t}{L} + n\right)^2\right] \right\}
\end{aligned}
\tag{9}
$$

It holds for $t \geq 2$. The first term corrects for the missing ($q = 0$) term in Eq (6). The remainder is a superposition of an infinite number of traveling Gaussian wave packets, initially excited in each replica and produced by the periodic boundary conditions. The time difference between the adjacent peaks is given as the acoustic traversal time τ_a. The widths of these peaks are increasing as $\sqrt{4(D+\Gamma)t}$ due to the diffusion of the tagged particle and the damping of sound modes. The relative magnitude of the peak separation and peak widths determines whether the negative plateau or the damped oscilations can be observed in specific time region. In particular, in the time interval, satisfying the inequalities

$$\sqrt{4(D+\Gamma)t} << c_0 t << L \ , \tag{10}$$

the 1-D VACF in Eq (6) shows very markedly a ***negative plateau***, given by the opposite of the ($q = 0$)-term, *i.e.* $\phi_{plat} = -(1-f)/N$. Here the spreading $\sqrt{4(D+\Gamma)t}$ of the wave packets is small compared to the propagation distance $c_0 t$ and simultaneously $t < \tau_a$. For times t much larger than the acoustic traversal time $\phi_N(t)$ approaches again zero.

To confirm the prediction, we carried out a simulation up to much larger times and obtained the result shown in Fig. 3. There is excellent agreement between

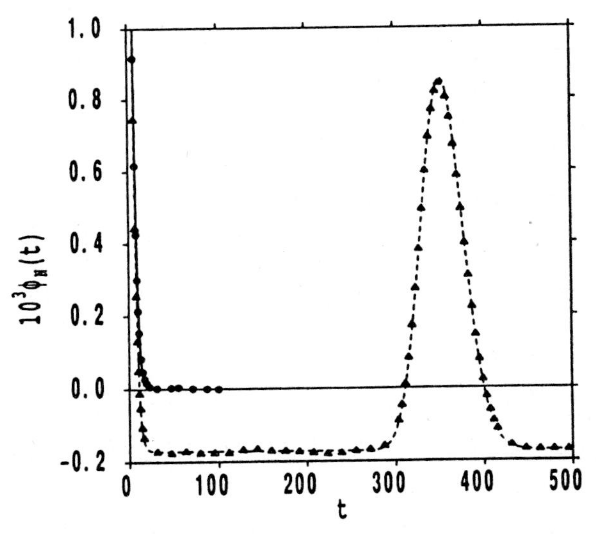

Fig. 3. The VACF in the 1-D model for $L = 500$ (▲) and $L = 10^5$ (•) at $f = 0.7$. The dashed curve shows the finite hydrodynamics prediction for the $L = 500$ system, Eq (6). The solid curve represents the extended mode coupling prediction for a infinite system, Eq (7).

simulations and theory. In the data for the $L = 500$ system there is a negative plateau for $25 < t < 275$ and another one for $t \geq 430$. They are seperated by a sharp peak, occurring at the acoustic traversal time $\tau_a = L/c_0 = 354$. This constant anticorrelation can be understood as follows. Since the tagged particle is given an initial velocity of **c** and the ensemble is prepared in such a way that the total momentum vanishes, each fluid particle has on the average a velocity $-\mathbf{c}/N$ yielding the anticorrelation $\phi_{plat} = -(1-f)c_0^2/N$, where the factor $(1-f)$ comes from the Fermi exclusion. In the time interval satisfying Eq (10), the fluid momentum (without the tagged particle) is essentially a constant of motion. Figure 3 also shows the simulation data for a system of $V = L = 10^5$ lattice sites at the same density. Here finite size effects are completely negligible and the simulations coincide with the result Eq (7) for the infinite system. Again, there is excellent agreement between simulations and the extended mode coupling theory.

To confirm the damped oscillations in Fig. 2 we performed a simulation for much larger times on a much smaller system of $L = 50$ with $\tau_a = 35.4$ in which the VACF is expected to give damped oscilations at much earlier times than in the $L = 500$ system. The mode coupling results, shown in Fig. 4, are again in excellent agreement with the computer simulations. The oscilations are due to

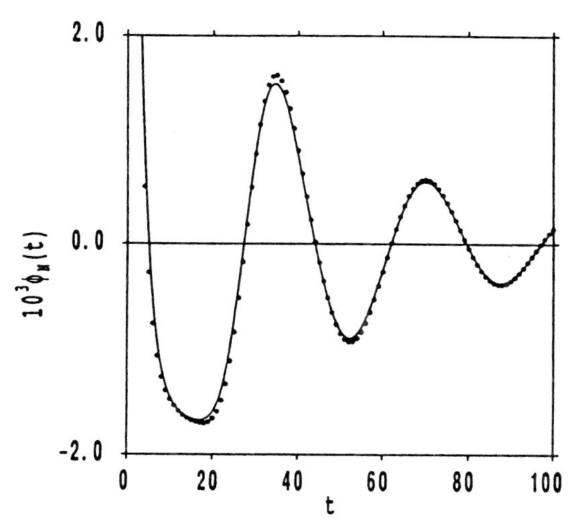

Fig. 4. The VACF in the 1-D model at $f = 0.7$, and $L = 50$. The solid curve represents the finite hydrodynamics prediction. The both results give damped oscilations.

interference effects of the tails of the diffusive wave packets which already have arrived in the reference region from its neighbouring replicas.

The negative plateau value $\phi_{plat} = -(1-f)/N$ may also be viewed as an effective finite size correction for short enough time intervals, such that all interactions of sound waves with their periodic images are absent ($t < \frac{1}{2}\tau_a$), where $\frac{1}{2}\tau_a$ is 177 and 18 in Fig. 1 ($L = 500$) and Fig. 4 ($L = 50$) respectively. Next we consider the *corrected* VACF at finite N, corrected for finite size effects, *i.e.*

$$\begin{aligned}\phi_{corr}(t) &= \phi_N(t) + (1-f)/N \\ &= ((1-f)/N)\sum_q \cos(c_0 qt)e^{-(D+\Gamma)q^2 t} \quad . \end{aligned} \tag{11}$$

In the thermodynamic limit the q-sum can be replaced by an integral. This leads to Eq (7), where ${}_1F_1$ reduces for $d = 1$ to an exponential,

$$\phi(t) = \frac{1-f}{bfc_0}\frac{1}{\sqrt{\pi t_h t}}\exp(-t/t_h) \quad . \tag{12}$$

Here $t_h = 4(D+\Gamma)/c_0^2$ denotes the crossover time from kinetic to hydrodynamic relaxation. This is shown in Fig. 5, where $\log[\phi_{corr}(t)]$ is plotted as a function of t at a density $f = 0.7$ with $t_h = 3.1$. For $t \leq 2t_0 \simeq 2$ there is an exponential

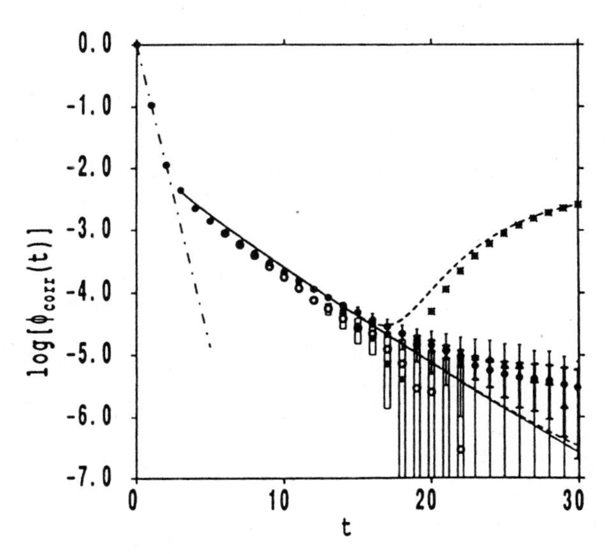

Fig. 5. Data collapse of corrected finite hydrodynamics results and M.D. data for $\phi_{corr}(t) = \phi_N(t) + (1-f)/N$ at $f = 0.7$ ($*$ for $L = 50$, ▲ for $L = 500$, o for $L = 2000$ and • for $V = 10^5$) with the infinite system result $\phi(t)$ (solid line; Eq (7)). The finite hydrodynamic predictions Eq (11) are shown as the dashed lines. The standard deviations are shown by thin vertical lines ($L = 500$), thick ones ($L = 10^5$) and vertical pillars ($L = 2000$). In the $L = 2000$ system the absolute errors (not all shown) for $t > 20$ are approximately 10^{-5}, whereas $\phi_{corr}^{MD}(t) < 10^{-7}$ for $t = 21,\ 23-30$.

decay with a kinetic relaxation rate that is correctly described by the Boltzmann equation. At $t = t_h \simeq 3.1$ there is a crossover to hydrodynamic decay bounded by an exponential with a slower decay rate $1/t_h$ (see Eq (12)). Indeed figure 5 shows the collapse of the corrected M.D. data, the corrected finite hydrodynamics result (11) and the infinite system result (7) in the appropriate time interval $t \leq 17$. For the smallest system ($L = 50$) the M.D. data (without error bars) and the finite hydrodynamics results should deviate from the infinite system curve for $t \geq \frac{1}{2}\tau_a$, as explained in Fig. 3. In the remaining data the error bars increase strongly for $t > 17$ because $\phi_N(t)$ becomes *negative* for $t \geq 12$ (M.D. 11), 15 (M.D. 14) and 26 (M.D. 84) for the system sizes $L = 500$, 2000 and 10^5 respectively. The quantity $\phi_{corr} = \phi_N + (1-f)/N$ vanishes exponentially fast and becomes small compared the absolute errors, which are of order 10^{-5} for all system sizes.

4 Discussion

Conclusions and outlook are summarized in a number of points:

(i) There is excellent agreement between the finite hydrodynamics results Eqs (6) and (9) for the VACF in 1-D CA-fluids and the computer simulation results over *more than five decades* in orders of magnitude.

(ii) The finite size correction Eq (11) permit us to collapse the M.D. data for all system sizes ($L = 50,\ 500,\ 2000,\ 10^5$) with the infinite system results in the appropriate time interval $t < \frac{1}{2}\tau_a$.

(iii) The short time kinetic theory result for $t \leq 2t_0$ (the mean free time $t_0 \sim 1/f$) combined with the finite hydrodynamics results for $t \geq 3t_0$ essentially provides a quantitative theory for the VACF in this 1-D CA-fluid. The crossover from collisional to hydrodynamic relaxation occurs at $t \simeq 2t_0$ without any complicated intermediate dynamics, such as the cage effect. This is probably related to the absence of any structure in the pair correlation function of this ideal Fermi gas.

(iv) The finite hydrodynamics result Eq (6) can also be applied to higher dimensional systems, as will be discussed extensively in Ref [8]. As an illustration we compare in Fig. 6 the theoretical results for the VACF $\phi_{corr} = \phi_N + (1-f)/N$, corrected for finite size effects, with the M.D. data of Ref [1-4] in the two-dimensional FHP-fluid for system size $V = L^2 = 50 \times 50$ at $f = 0.1$ and $V = 100 \times 100$ at $f = 0.75$. The minimum in the lower density curve ($f = 0.1$) at $t = \frac{1}{2}\tau_a \simeq 37$ is consistent with the acoustic traversal time $\tau_a = L/c_0$ with $c_0 = \sqrt{3/7}$ and completely analogous to the minimum in Fig. 4 at $t = \frac{1}{2}\tau_a = 18$.

(v) We compare the M.D. data for the 2-D FHP-III model with the restricted mode coupling results (only shear modes) of Ref [1-4] and with the extended mode coupling results (shear modes and sound modes) for finite (short dashes) and infinite system (solid line). The plot illustrates the importance (a) of including sound modes to extend the validity of mode coupling theory to shorter times, (b) of finite hydrodynamics for $t > \frac{1}{2}\tau_a$ in small systems to account for interference effects of the periodic replica systems, and (c) of the finite size correction $(1-f)/N$ to collapse small and large system results for $t < \frac{1}{2}\tau_a$.

(vi) In the quasi 3-D FCHC model the system is actually a 4-D slab, one lattice spacing wide in the fourth dimension. How to account for finite size effects in this geometry is under investigation. In general the differences between restricted and extended mode coupling theory and between finite and infinite systems are quite small. The finite size corrections of Eq (11) is typically of order 10^{-6} and can be neglected.

Acknowlegements

T.N thanks the Institute of Theoretical Physics of the University of Utrecht for its kind hospitality during his stay. The work of the FOM Institute is part of the scientific program of FOM and is supported by 'Nederlandse Organisatie voor Wetenshappelijk Onderzoek' (NWO). Computer time on the NEC-SX2 at NLR was made available through a grant by NFS (Nationaal Fonds Supercomputers).

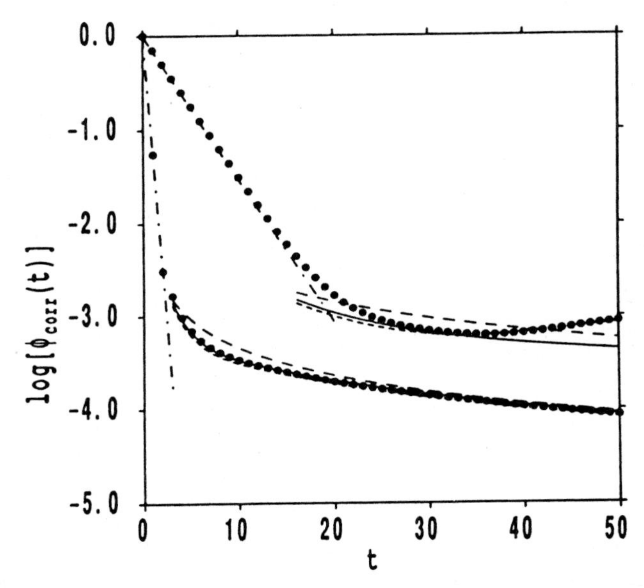

Fig. 6. The logarithm of the $\phi_{corr}(t)$ versus time in 2-D FHP-III model for system size $V = 50 \times 50$ at density $f = 0.1$ and for $V = 100 \times 100$ at $f = 0.75$. The symbols •'s represent the results of computer simulations. The solid, dotted and dashed curves denote results respectively obtained from the (extended) infinite mode coupling theory, the finite hydrodynamics and the leading t^{-1}–tails. The short time M.D. data agree with the Boltzmann approximation.

References

1. Frenkel, D. and Ernst, M.H., *Phys. Rev. Lett.* 63, 2165(1989)
2. van der Hoef, M. and Frenkel, D., *Phys. Rev.* A41, 4277(1990)
3. van der Hoef, M. and Frenkel, D., in *Proceeding on Lattice Gas Methods for PDE's, Los Alamos, Sept., 1989,* Doolen, G.D., Ed., *Physica* D, Nov. 1990
4. Ernst, M.H., in *Proceeding on Lattice Gas Methods for PDE's, Los Alamos, Sept., 1989,* Doolen, G.D., Ed., *Physica* D, Nov. 1990
5. Ernst, M. H. and Dufty, J. W., *J. Stat. Phys.* 58, 57(1990)
6. Naitoh T., Ernst M.H. and Dufty J.W., *Phys. Rev.* A15 to appear
7. Erpenbeck, J.J. and Wood, W.W., *Phys. Rev.* A26, 1648(1982)
8. Naitoh, T., Ernst, M.H., Frenkel, D. and van der Hoef, M., in preparation
9. Green, M.S., *Phys. Rev.* 119 829(1960)
10. Lebowitz, J.L. and Percus, J.K., *Phys. Rev.* 122, 1075(1961)
    Lebowitz, J.L., Percus, J.K. and Verlet, L., *ibid.* 153, 250(1967)
11. d'Humières, D., Lallemand, P. and Qian, Y.H., *C.R. Acad. Sci. Paris* II 308,585 (1988)
12. Ziman, J.M., in *Principles of the Theory of Solids*, P. 40, Eq (2.34), (Cambridge Univ. Press, 1972)

This article was processed using the LaTeX macro package with ICM style

Hydrodynamic Behaviour of the Lattice Boltzmann Equation

S. Succi[1], *R. Benzi*[2], *M. Vergassola*[3] *and A. Cancelliere*[4]

[1] IBM European Center for Scientific and Engineering Computing, via Giorgione 159, I-00147 Roma, Italy
[2] Dipartimento di Fisica, Universitá 'Tor Vergata', via E. Carnevale, I-00173 Roma, Italy.
[3] Dipartimento di Fisica, Universitá 'La Sapienza', P.le A. Moro 2, I-00185 Roma, Italy
[4] Universitá di Catania, Piazza dell' Universitá, I-95124 Catania, Italy

October 6, 1990

Abstract

The hydrodynamic behaviour of the Lattice Boltzmann Equation is discussed. It is shown that the projection of the macrodynamic equations onto the eigenvectors of the collision operator allows a clear analysis of the convergence of the model to the Navier-Stokes equations. A natural distinction between Hydrodynamic and non-hydrodynamic variables arises. The effect of the discreteness of the lattice is discussed in the case of a simple deterministic Lattice Boltzmann Equation having a macroscopic diffusive behaviour.

In the last few years, the Lattice Boltzmann Equation (LBE) has been introduced as a new computational approach to the study of two and three dimensional fluid flows[1,2,3]. LBE can be obtained under the assumption of molecular chaos, from the discrete boolean dynamics of a Lattice Gas Automaton. Recently[4], it has been shown that one can consider LBE as a starting model of fluid flows, independently of the detailed collision rules describing the Lattice Gas Automaton. In particular, by expanding LBE up to second order in the velocity field, one can obtain an efficient algorithm for fluid dynamics[5]. Here we will discuss the hydrodynamic limit of LBE, introducing also a new Lattice Boltzmann Equation having a macroscopic diffusive behaviour which allows to analyze the effects of the discreteness of the lattice.

The starting point of our analysis is the class of LBE defined in [4], which governs the evolution of the mean particle population N_i in the i−th state by

$$N_i(\vec{x}+\vec{c}_i,t+1)-N_i(\vec{x},t)=\Omega_{ij}(N_j-N_j^{eq})\quad i=1,\ldots,b. \tag{1}$$

Here b is the number of link per site, $\vec{x}$ runs over the lattice sites, $\vec{c}_i$ are the possible velocities, $c=|\vec{c}_i|$, D is the dimension of the lattice and

$$\left\{\begin{array}{l} N_i^{eq}=\frac{\rho}{b}\left(1+\frac{D}{c^2}v_\alpha c_{i\alpha}+\frac{D^2}{2c^4}\frac{b-2\rho}{b-\rho}Q_{i\alpha\beta}v_\alpha v_\beta\right), \\ Q_{i\alpha\beta}=c_{i\alpha}c_{i\beta}-\frac{c^2}{D}\delta_{\alpha\beta}, \qquad \rho=\sum_{i=1}^{b}N_i, \qquad \rho v_\alpha=\sum_{i=1}^{b}N_ic_{i\alpha}. \end{array}\right\} \tag{2}$$

The matrix Ω is not defined by scattering rules. The general element Ω_{ij} is determined by the symmetry requirement that it depends only on the angle between directions $\vec{c}_i$ and $\vec{c}_j$ and by imposing the conservation of mass and momentum. The matrix Ω is symmetric and cyclic. Its eigenvectors are orthogonal, do not depend on the matrix elements and it is easy to compute its eigenvalues[7]. In the case of the four-dimensional face-centered-hypercubic (FCHC) lattice[8], denoting by a_θ the matrix elements Ω_{ij} such that $\vec{c}_i\cdot\vec{c}_j=c^2\cos(\theta)$, the non-zero eigenvalues are

$$\lambda=a_0-2a_{90}+a_{180}\,,\ \sigma=\frac{3}{2}(a_0-a_{180})\,,\ \tau=\frac{3}{2}(a_0+6a_{90}+a_{180}). \tag{3}$$

with multiplicities 9, 8 and 2 respectively. From the four-dimensional lattice one can obtain a $2D$ lattice by projecting on the $x-y$ plane. In this projection some directions generate the same $2D$ vector and only the sum of their populations is relevant to the calculation of two-dimensional fields. We can therefore assume that the N_i corresponding to the same $2D$ vector are degenerate. Because the dynamics preserves these degeneracies, it is possible to define a 9×9 reduced scattering matrix $\mathcal{C}$. Remark that the introduction of the matrix $\mathcal{C}$ is very interesting also practically. In fact, there is no need to refer to the complete model and one can work directly with the reduced model with just nine populations. The reduction of computational time and memory is huge.

Let us come back to the theoretical analysis. The matrix $\mathcal{C}$ is not symmetric and cyclic. However, $\mathcal{C}$ has the same set of eigenvalues of Ω $(0, \lambda, \sigma, \tau)$ with multiplicities $3, 3, 2, 1$ respectively. One can define a basis of orthogonal eigenvectors by choosing the following definition of scalar product

$$\vec{A} \cdot \vec{B} = \sum_{i=1}^{9} p_i A_i B_i \quad p_i = \begin{cases} 1 & \text{for diagonal directions} \\ 4 & \text{otherwise} \end{cases} \tag{4}$$

It follows that N_i can be decomposed as

$$N_i(\vec{x}, t) = \sum_{n=1}^{9} A_i^{(n)} \phi^{(n)}(\vec{x}, t). \tag{5}$$

It is readily checked that

$$A_i^{(1)} = 1_i \equiv (1, \ldots, 1), \quad A_i^{(2)} = c_{ix}, \quad A_i^{(3)} = c_{iy}. \tag{6}$$

and the corresponding eigenvalue is zero. It is possible to identify $\phi^{(1)}$ as ρ and $\phi^{(2)}$, $\phi^{(3)}$ as the $x - y$ projection of $\vec{J} = \rho\vec{v}$. On the other hand, after some algebra, one finds that

$$A_i^{(7)} = r_i c_{ix}, \quad A_i^{(8)} = r_i c_{iy}, \quad A_i^{(9)} = (1, -2, 1, -2, 1, -2, 1, -2, -2) = r_i. \tag{7}$$

with eigenvalues σ, σ, τ respectively. The eigenvectors $\phi^{(7)}$ and $\phi^{(8)}$ should be considered as the $x - y$ projection of a vector field $\vec{\eta}$. This vector field has the physical meaning of a density current, akin to $\vec{J}$, whose density field is $\phi^{(9)}$, hereafter denoted by μ. Both $\vec{\eta}$ and μ have no immediate hydrodynamical meaning and we shall refer to them as ghost fields. Although $A_i^{(5)}$ is simply given by Q_{ixy}, $A_i^{(4)}$ and $A_i^{(6)}$ are given by a linear combination of Q_{ixx} and Q_{iyy}:

$$A_i^{(4)} = Q_{ixx}\,, \; A_i^{(6)} = Q_{iyy} + \frac{1}{3} Q_{ixx}. \tag{8}$$

Hereafter we find it more convenient to use the tensor $Q_{i\alpha\beta}$ instead of $A_i^{(4)}$, $A_i^{(5)}$, $A_i^{(6)}$. In this case we loose the orthogonality of the basis but, as we shall see below, the advantage is that the definition of the stress tensor $S_{\alpha\beta}$ becomes straightforward: $S_{\alpha\beta} = \sum_i p_i N_i Q_{i\alpha\beta}$. The corresponding eigenvalue is λ.

In order to study the macroscopic behaviour of LBE, it is convenient to use the multiscale expansion discussed in [1]. Following [6], by keeping only first order terms in the multiscale expansion, we obtain the differential equations:

$$\partial_t N_i + (\vec{c}_i \cdot \vec{\partial}) N_i = C_{ij}(N_j - N_j^{eq}). \tag{9}$$

Corrections to these equations will be discussed later. By using the properties of the basis listed above, we can project (9) onto the nine eigenvectors obtaining the following set of non linear partial differential equations:

$$\left\{ \begin{array}{cc} \partial_t \rho + \partial_\alpha J_\alpha = 0, & \partial_t \mu + \partial_\alpha \eta_\alpha = \tau \mu, \\ \partial_t J_\alpha + \partial_\alpha (\frac{\rho}{2}) + \partial_\beta S_{\alpha\beta} = 0, & \partial_t \eta_\alpha + \partial_\alpha (\frac{\mu}{2}) + \partial_\beta T_{\alpha\beta} = \sigma \eta_\alpha, \\ \multicolumn{2}{c}{\partial_t S_{\alpha\beta} + \frac{H_{\alpha\beta}}{3} + G_{\alpha\beta} = \lambda (S_{\alpha\beta} - S_{\alpha\beta}^{eq}) \text{ with } S_{\alpha\beta}^{eq} = \rho g(\rho)(v_\alpha v_\beta - \frac{v^2}{4}\delta_{\alpha\beta}).} \end{array} \right\} \tag{10}$$

Here

$$\left\{ \begin{array}{l} T_{\alpha\beta} = \sum_i p_i r_i N_i Q_{i\alpha\beta},\ g(\rho) = \frac{2}{3}\frac{1-\rho/12}{1-\rho/24},\ G_{\alpha\beta} = R_{\alpha\beta\gamma\nu}\partial_\gamma \eta_\nu, \\ H_{\alpha\beta} = \partial_\alpha J_\beta + \partial_\beta J_\alpha - \frac{\partial_\gamma J_\gamma}{2}\delta_{\alpha\beta},\ R_{\alpha\beta\gamma\nu} \equiv \frac{\sum_i p_i r_i c_{i\alpha} c_{i\beta} c_{i\gamma} c_{i\nu}}{24}. \end{array} \right\} \tag{11}$$

Remark that the left hand side of equations (10) for the ghost fields and the hydrodynamical fields have the same formal structure: one can then switch from one block to another by simply letting $p_i \leftrightarrow p_i r_i$. This reflects the basic difference between the two kinds of variables. By equation (11) one obtains:

$$T_{xx} = S_{xx} - \frac{3}{2}Tr(S) \qquad T_{yy} = S_{yy} - \frac{3}{2}Tr(S) \qquad T_{xy} = -2S_{xy} \tag{12}$$

As we shall see below, under certain conditions S is the isotropic NS stress tensor. On the contrary, the tensor $T_{\alpha\beta}$ is not isotropic, because of the factor -2 in the definition of the T_{xy} component.

By using the same kind of calculations as in $2D$, one can obtain the macroscopic equations in three dimensions. It is then possible to show that the equations are structured as two blocks of variables evolving in the same way as in $2D$. The considerations about the two dimensional case apply also to $3D$.

Three conditions are necessary in order to get NS equations. The first is that ghost fields are much smaller than hydrodynamical fields; the second is the condition of incompressibility $M \ll 1$, where M is the Mach number; the third is the validity of the adiabatic approximation, namely the possibility of neglecting the time derivative in the evolution equation of the stress tensor. These conditions imply that $S_{\alpha\beta} = S^{eq}_{\alpha\beta} + \frac{1}{3\lambda} H_{\alpha\beta}$ and, by substituting this expression in the equation for $\vec{J}$, we obtain NS equations with a viscosity $1/3|\lambda|$, apart from the usual constant factor $g(\rho)$.

The first condition is strictly related to the issue of higher order terms of the multiscale expansion used to obtain the continuous equations from the discrete model. From equation (10) one obtains

$$S \sim v^2 + \epsilon v \qquad \eta \sim \epsilon v^2 + \epsilon^2 v \qquad \mu \sim O(\epsilon^2). \tag{13}$$

Because the corrections coming from higher order terms in the multiscale expansion are $O(\epsilon^2)$, one can argue that, as $\epsilon \to 0$, the rates $\frac{\eta}{S}$ and $\frac{\mu}{S}$ go to zero. The activity of ghost fields is confined to the short scale dynamics, where the local Knudsen number is $O(1)$ and no Chapman-Enskog expansion is available. On the other hand, by substituting the adiabatic expression of S in the equation for $\vec{J}$, the advective term is $O(\epsilon v^2)$, while the dissipative one is $O(\epsilon^2 v)$. It is therefore necessary to include second order terms of the multiscale expansion. The calculation is the same as in [1] and the final result is the inclusion of the propagation contribution in the expression of viscosity.

The second condition $M \ll 1$ corresponds to the usual requirement that velocities are much smaller than the speed of sound.

The third condition required to obtain NS equations is the validity of the adiabatic approximation. For a simplified equation of the form

$$\partial_t x(t) + f(t) = -\varsigma x(t) \tag{14}$$

where $\varsigma > 0$ and $f(t)$ is a forcing term whose characteristic time of variation is τ, the adiabatic approximation works in the limit $\varsigma \gg 1/\tau$. We are requiring that non-equilibrium dynamics is quick enough to suppose that the stress tensor is always in equilibrium with respect to the forcing term. In the case of (10), it is not trivial to identify the conditions of validity of the hypothesis, because the forcing depends on the fields themselves. To clarify this point, let us consider the relaxation of a given initial configuration as a function of λ, σ and τ. We disregard non linear terms which do not contribute to dissipation for an incompressible fluid. In this limit, in a real fluid, non-equilibrium components are rapidly slaved, while the current decays exponentially at a rate proportional to the viscosity. As already discussed, in the macroscopic limit ghost fields contributes a small correction and will not be considered. We look for the solution in terms of a Fourier series. Thanks to isotropy, one can consider the case $k_y = 0$ and the substitution $k_x \mapsto k$ is sufficient to get the eigenvalues in full generality. There are two different kinds of behaviour. In the limit $|\lambda| \gg k$, at leading order in $k/|\lambda|$, one gets:

$$\left\{ \begin{array}{c} J_y(t) = J_y(0)\exp(\omega_1 t) + -i\frac{k}{\lambda}S_{xy}(0)\{\exp(\omega_2 t) - \exp(\omega_1 t)\} \\ S_{xy}(t) = S_{xy}(0)\exp(\omega_2 t) + \frac{ik^2}{3\lambda}J_y(0)\exp(\frac{k^2 t}{3\lambda}), \end{array} \right\} \tag{15}$$

where $\omega_1 = k^2/3\lambda = -\nu k^2$ and $\omega_2 = \lambda - k^2/3\lambda$. In this case the results are the same that one would obtain by using the adiabatic approximation, i.e. the behaviour is hydrodynamical. In the opposite limit, one can show that the relaxation time is very long and the non equilibrium field S is not slaved to the hydrodynamic field J: there is no time scale separation between the approach to thermodynamical equilibrium and the hydrodynamical decay. The condition $|\lambda| \gg k$ is therefore necessary to get NS equations. This is in agreement with the experimental observation that hydrodynamic behaviour is attained within a few lattice spacing. We observe that non-equilibrium fluctuations with a characteristic spatial scale $1/k$ have a relaxation time of order $1/k$, as it should be, because the speed of sound is order one.

From this condition one can see that LBE converges to NS equation better and better as the viscosity decreases, i.e. in the limit of high Reynolds numbers. On the other hand, in this limit short-scales are excited. The discreteness of the lattice becomes of relevance. In order to clarify this point, it is useful to consider a simplified one-dimensional deterministic model which exhibits purely diffusive behaviour on a macroscopic scale.

Let us consider a one-dimensional lattice with two states per node representing unit-speed particles moving rightwards and leftwards respectively. The collision matrix is defined by the mass conservation. The parameter a is the matrix element coupling the two directions. The same procedure used previously can be repeated in this case. It is then possible to show that the macroscopic behaviour of the model is well represented by the following set of equations

$$\begin{aligned} \partial_t \rho + \partial_x J + \frac{1}{2}\partial_x \partial_x \rho = 0 \\ \partial_t J + \partial_x \rho = -2aJ \equiv \lambda J. \end{aligned} \tag{16}$$

The diffusion coefficient in the macroscopic limit is given by

$$D(\lambda) = -\frac{1}{\lambda} - \frac{1}{2} \qquad -2 < \lambda < 0. \tag{17}$$

Remark that $D(\lambda)$ depends on the eigenvalue λ in the same way as viscosity in hydrodynamics. In particular, in the limit $\lambda \to -2$, the diffusion coefficient vanishes.

The model is interesting because it is amenable to an exact solution in its fully discrete form, with no need of any multiscale expansion. The equation is linear so that solutions are most conveniently expressed in terms of a plane-wave expansion. It is easy to obtain the dispersion relation. One of the eigenvalues describes a pure relaxation, while the other mode is associated with diffusive behavior. The frequency of the diffusive mode can be recasted in the following form

$$\omega_+ = \frac{D(\lambda)}{Q(\lambda, \theta)}. \tag{18}$$

Here $D(\lambda)$ is the macroscopic limit of the diffusion coefficient, θ is the wavenumber normalized to the lattice spacing and

$$Q(\lambda, \theta) = -\frac{\theta^2}{2}\frac{1-\lambda}{\lambda} 1/\log\left[(1-\lambda)\cos\theta + \sqrt{\lambda^2 - (1-\lambda)^2 \sin^2\theta}\right], \tag{19}$$

is a function which embodies the effects of the lattice discreteness. As one can check by direct inspection, $Q(\lambda, \theta) \to 1$ as $\theta \to 0$ for any value of λ. It is now of interest to examine the effect of the lattice on the dynamics of the short-scales as a function of λ. A series of curves for different values of λ are shown in figure 1. From this figure, we see that

a) For small value of λ, the short scales are overdamped.

b) As λ approaches -2, the short scales do not relax.

c) The region of convergence to the macroscopic value extends as $|\lambda|$ grows when $|\lambda|$ has a value far enough from -2. In the neighbourhood of -2 the tendency is reversed and the transition to the non-dissipative behaviour at high-k tends to become much more abrupt.

We argue that this behaviour has a certain generality in the sense that also in hydrodynamics the scheme becomes unsteady in the limit $\lambda \rightarrow -2$. When transferred to the context of the hydrodynamic LBE, these considerations suggest that by pushing the viscosity too low, undamped short-scales are excited. The consequence is that if the resolution is not sufficient, in the limit of small viscosity, the LBE scheme is prone to numerical instabilities. It is worthwhile to stress that the diffusion model can be simply modified to obtain a scheme which in the macroscopic limit reproduces the behaviour of the Burgers equation. In fact it is sufficient to introduce a term quadratic in ρ in the expression of the equilibrium population [9].

In conclusion, we have shown that, by using a suitable projection on the eigenvectors of the collision matrix, one can distinguish between hydrodynamical and non-hydrodynamical variables. This analysis allows to discuss the hydrodynamic limit of the Lattice Boltzmann Equation and to show that the model is a hyperbolic approximation, in the sense of finite differences, of the Navier-Stokes equations. We have also considered a simple model having macroscopic diffusive behaviour which can be solved in its fully discrete form. It suggests that when the Reynolds number is high and the resolution is not sufficient the hydrodynamic Lattice Boltzmann Equation tends to become unstable.

References.

1. U. Frisch, D. d'Humieres, B. Hasslacher, P. Lallemand, Y. Pomeau, J.P. Rivet, *Complex Systems* **1**, 649 (1987).
2. G. Mc Namara, G. Zanetti, *Phys. Rev. Lett.* **61**, 20 (1988).
3. F. Higuera, J. Jimenez, *Europhys. Lett.* **9**, 663 (1989)
4. F. Higuera, S. Succi, R. Benzi, *Europhys. Lett.* **9**, 345 (1989)
5. R. Benzi, S. Succi, *J. Phys.* **A23**, L1 (1990)
6. U. Frisch in, *Lattice Gas Methods for PDE's* , Los Alamos, September 6-9, 1989 (ed. G. Doolen) to be published in Physica D
7. S. Wolfram, *J. Stat. Phys.* **45**, 471 (1986).
8. D. d'Humieres, P. Lallemand, U. Frisch, *Europhys. Lett.* **2**, 291 (1986).
9. R. Benzi, S. Succi, M. Vergassola, *Physics Report*, in preparation.

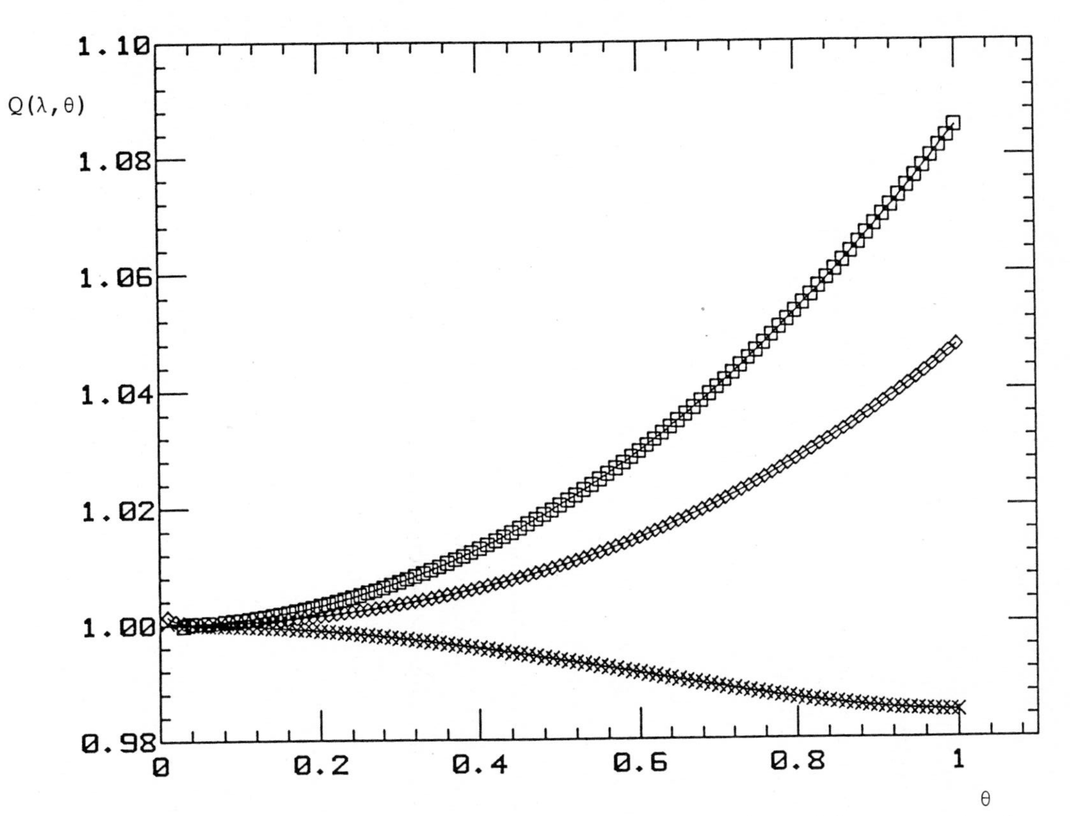

Fig. 1 The plot of $Q(\lambda, \theta)$ vs. θ for three different values of the parameter λ is shown. The curves (the order is from the bottom to the top) refer to the values $\lambda = 0.6$, $\lambda = 0.7$ and $\lambda = 0.9$.

Modified Lattice Gas Method For High Reynolds Number Incompressible Flows

Francisco J. Higuera

E.T.S.I. Aeronáuticos. Universidad Politécnica de Madrid. Pza. Cardenal Cisneros 3, 28040 Madrid. Spain

1 Introduction

Among the difficulties encountered in the application of the lattice gas method to incompressible fluid dynamics are those associated to dealing with the incompressibility condition, which are shared by many classical numerical methods. Since the mass and momentum conservation equations for a lattice gas are evolution equations, the incompressibility constraint $\nabla \cdot \mathbf{v} = 0$ holds only asymptotically, for very small Mach numbers and when no acoustic waves are excited in the gas. This is a rather precarious situations when the main goal is to describe incompressible flows. Since a Courant-Friedrick-Levy stability condition is built into the automaton rules, the time step, limited by the fast acoustic waves, is of the order of the lattice pitch (h) divided by the velocity of the particles (c), while the characteristic time for the evolution of the gas (T) is much longer, of the order of the characteristic length of the problem (L) divided by the characteristic velocity of the flow(U). Hence, the number of time steps scales like $(L/U)/(h/c) \sim (L/h)/M$, where M is the Mach number. The factor $1/M \gg 1$ would disappear for a really incompressible fluid, without acoustic modes, but for such a fluid the condition $\nabla \cdot \mathbf{v} = 0$ must be enforced explicitly, which makes the problem strongly nonlocal. As the price of locality ($1/M$) may be very high, it pays facing the nonlocal elliptic problem directly. In this paper we discuss modifications of the lattice gas method, inspired in iterative Poisson solvers, that partially bring the elliptic problem back to the realm of lattice gases.

2 Model Without Acoustic Modes

Let us consider first a continuous system $N_i(\mathbf{x}, t)$, $\pi(\mathbf{x}, t)$, with $i = 1, \ldots, b_N$ and $\mathbf{x} \in \mathrm{I\!R}^D$, whose evolution obeys the equations

$$\frac{\partial N_i}{\partial t} + \mathbf{c}_i \cdot \nabla N_i = \Omega_i - \gamma \mathbf{c}_i \cdot \nabla \pi, \ , \qquad (1)$$

$$\nabla \cdot \mathbf{v} = 0 \ , \qquad (2)$$

with $\gamma = D/(c^2 b_N)$, $c = |\mathbf{c}_i|$, and $\mathbf{v} = \sum_i^{b_N} N_i \mathbf{c}_i$. Here the vectors $\mathbf{c}_i$ satisfy the relations

$$\sum_i c_{i\alpha} c_{i\beta} = \frac{c^2 b_N}{D} \delta_{\alpha\beta}$$

$$\sum_i c_{i\alpha} c_{i\beta} Q_{i\gamma\delta} = \frac{c^4 b_N}{D(D+2)} (\delta_{\alpha\gamma}\delta_{\beta\delta} + \delta_{\alpha\delta}\delta_{\beta\gamma} - \frac{2}{D}\delta_{\alpha\beta}\delta_{\gamma\delta}) \ ,$$

where the greek letters denote cartesian components in the D-dimensional space, and

$$Q_{i\alpha\beta} = c_{i\alpha} c_{i\beta} - \frac{c^2}{D} \delta_{\alpha\beta} \ .$$

The operator Ω has the form

$$\Omega_i = A_{ij} \left(N_j - N_j{}^{eq} \right) \ , \tag{3}$$

with A_{ij} symmetric negative semidefinite and

$$N_i{}^{eq} = \frac{1}{b_N} \left(1 + \frac{D}{c^2} c_{i\alpha} v_\alpha + G Q_{i\alpha\beta} v_\alpha v_\beta \right) , \quad G = \frac{D(D+2)}{2c^4} \ .$$

Equation (1) is the analogous of the Boltzmann equation for a gas with discrete velocities $\mathbf{c}_i$, see [1] for further details. However, we'll assume here that the collision operator Ω conserves only the momentum but not the mass; i.e. the null space of A is spanned by the D vectors $(c_i)_\alpha$, in a b_M dimensional vector space, with $\alpha = 1, \ldots, D$. $I = (1, \ldots, 1)$ is an eigenvector of A with associated eigenvalue $-\alpha < 0$, and $Q_{i\alpha\beta}$, $\alpha, \beta = 1, \ldots, D$, are all of them eigenvectors associated to the same eigenvalue $-\lambda < 0$.

The variable π plays the role of a Lagrange multiplier enforcing the condition (2). Taking the divergence of (1) and using (2) we find $\nabla^2 \pi = -\Phi$, where

$$\Phi = \partial_\alpha \partial_\beta \left(\sum_i Q_{i\alpha\beta} N_i \right) + \frac{c^2}{D} \nabla^2 \left(\sum_i N_i \right) \ . \tag{4}$$

In what follows the time and length units will be supposed to be such that c and the eigenvalues of A are quantities of the order of the unity. More precise definitions are postponed until Sect. 4.

A Chapman-Enskog multiscale expansion in the hydrodynamical limit of slowly varying field ($\partial/\partial t = O(1/T) \ll 1$, $\partial/\partial x_\alpha = O(1/L) \ll 1$) shows that $\mathbf{v}$ satisfies the Navier-Stokes equations, except for a spurious term commented on below, and π is related to the pressure. Since (2) is verified exactly, the large scale behaviour of this system mimics that of an incompressible fluid, without acoustic modes. The asymptotic analysis is similar to those of references [1,2], with obvious changes to account for the reduced set of conservation laws, and only the results are quoted. Note that the velocity of the flow, or the Mach number, is not assumed to be small in what follows, except where indicated explicitly.

The Euler equations (corresponding to the projection of (1) onto the null space of A with $N = N^{eq}$) are

$$\frac{\partial \mathbf{v}}{\partial t} + \mathbf{v} \cdot \nabla \mathbf{v} = -\nabla \left(\pi - \frac{v^2}{D} \right) . \tag{5}$$

The nonequilibrium part of the distribution function in the Chapman-Enskog expansion $N = N^{eq} + N^{ne}$, with $N^{ne} = O(1/L)$, is

$$N_i{}^{ne} = (\Omega^{-1})_{ij} \left\{ \frac{\partial N_j{}^{eq}}{\partial t} + \mathbf{c}_j \cdot \nabla N_j{}^{eq} \right\}_\perp =$$
$$-\frac{1}{\lambda} \frac{Q_{i\alpha\beta}}{b_N} \left(G \frac{\partial v_\alpha v_\beta}{\partial t} + \frac{D}{c^2} \partial_\alpha v_\beta \right) + \frac{G}{b_N} (\Omega^{-1})_{ij} Q_{j\alpha\beta} c_{j\gamma} \partial_\gamma (v_\alpha v_\beta) , \tag{6}$$

where $\perp$ means component orthogonal to the null space of A. The pressure term $\gamma \mathbf{c}_i \cdot \nabla \pi$ belongs to the null space of A and therefore does not give any contribution to N^{ne}. The last term on the right side is orthogonal to I and to $Q_{i\alpha\beta}$.

The Navier-Stokes equations (corresponding to the projection of (1) onto the null space of A with $N = N^{eq} + N^{ne}$) are

$$\frac{\partial \mathbf{v}}{\partial t} + \mathbf{v} \cdot \nabla \mathbf{v} = -\nabla \left(\pi - \frac{v^2}{D} \right) + \frac{1}{\lambda} \frac{c^2}{D(D+2)} \nabla^2 \mathbf{v}$$
$$+ \frac{1}{\lambda} \frac{\partial}{\partial t} \left(\nabla \cdot (\mathbf{v}\ \mathbf{v}) - \frac{\nabla v^2}{D} \right) . \tag{7}$$

Here the time derivative on the right side must be eliminated using the Euler equations: it is displayed in this way for the sake of brevity. The term $-\nabla(\partial v^2/\partial t)/\lambda D$ is to be included in the definition of the pressure, because it is a gradient. On the contrary, $1/\lambda \partial \nabla \cdot (\mathbf{v}\ \mathbf{v})/\partial t$ is a spurious term of order

$$\frac{1}{\lambda} \frac{\partial}{\partial t} \nabla \cdot (\mathbf{v}\ \mathbf{v}) = O(\frac{U^2}{LT}) = O(\frac{U^3}{L^2}) = U^2 O(\frac{1}{\lambda} \nabla^2 \mathbf{v}) .$$

Hence $U^2 \ll 1$ is formally required to get rid of this spurious term. Note, however, that it is always much smaller than the inertia terms whatever the order of the velocity.

Finally, carrying $N = N^{eq} + N^{ne}$ to the Poisson equation for π, we find

$$\nabla^2 \pi = -\left(1 - \frac{1}{\lambda} \frac{\partial}{\partial t} \right) \nabla \cdot \left(\nabla \cdot (\mathbf{v}\ \mathbf{v}) - \frac{\nabla v^2}{D} \right) . \tag{8}$$

On eliminating $\partial \mathbf{v}/\partial t$ the pressure reappears on the right side. This means that the operator acting on π is not exactly ∇^2, but rather $\nabla^2 + \mathcal{L}$, with

$$\mathcal{L}\pi = \frac{-1}{\lambda D} \left\{ (D-1) \mathbf{v} \cdot \nabla (\nabla^2 \pi) + (D-2)(\nabla \mathbf{v} : \nabla\nabla\pi) - 2(\nabla^2 \mathbf{v}) \cdot (\nabla \pi) \right\} .$$

However, since $O(\mathcal{L}\pi) = (U/L) O(\nabla^2 \pi)$, $\mathcal{L}\pi$ can be safely neglected.

3 Two-Gas Model

The model introduced in the previous section mimics the dynamics of an incompressible fluid but it can not be reduced to a lattice gas: nonlocal operations are involved because Eqs. (1,2) are elliptic. Let us consider an alternative two-gas model with populations $N_i(\mathbf{x},t)$, $i = 1,\ldots,b_N$, for the primary gas, and $M_j(\mathbf{x},t)$, $j = 1,\ldots,b_M$, for a secondary gas, satisfying

$$\frac{\partial N_i}{\partial t} + \mathbf{c}_i \cdot \nabla N_i = \Omega_i - \gamma \mathbf{c}_i \cdot \nabla \pi \tag{9}$$

$$\frac{\partial M_j}{\partial t} + \mathbf{b}_j \cdot \nabla M_j = \overline{\Omega}_j + \beta \Phi \ , \tag{10}$$

where, now, $\pi = \sum_j M_j$ is the mean population of the M gas; Φ is given by Eq. (4); $\beta = b/(b_M D)$, with $b = |\mathbf{b}_j|$; and the vectors $\mathbf{b}_j$ satisfy

$$\sum_j b_{j\alpha} b_{j\beta} = \frac{b^2 b_M}{D} \delta_{\alpha\beta} \ .$$

The collision operator for the gas M is

$$\overline{\Omega}_j = b\left(-M_j + \frac{1}{b_M}\sum_k M_k\right) \ . \tag{11}$$

Hence, $I = (1,\ldots,1)$ is an eigenvector of $\overline{\Omega}$ with eigenvalue zero and each vector orthogonal to I is an eigenvector with eigenvalue $-b$; this operator conserves only the mass. The equilibrium populations of the gas M, leaving aside its interaction with the other gas, are $M_i{}^{eq} = \pi/b_M$.

The basic idea underlying this model is that the effects of convection and diffusion of the vorticity on one side, and of enforcing the incompressibility condition on the other (through the appropriate pressure forces) are different tasks involving very disparate time scales for low Mach number gas flows, and each task can be assigned to a different gas. Here gas M in intended as an iterative Poisson solver and, for this to be the case, its velocities must be large enough. In fact, the following scales are involved in the dynamics of the system:

$t_1 = O(1/b)$	: Collision (relaxation) time for gas M.
$t_2 = O(1/c) = O(1)$	: Collision (relaxation) time for gas N.
$t_3 = O(L^2/b)$	: Diffusion time for gas M (see below).
$t_4 = O(L/U)$	: Convection time for gas N.

A trivial Chapman-Enskog expansion for the gas M leads to

$$M_i{}^{ne} = \left(\overline{\Omega}^{-1}\right)_{ij}\left\{\frac{\partial \pi/b_M}{\partial t} + \mathbf{b}_j \cdot \nabla\pi/b_M\right\}_{\perp} = -\frac{b_i}{b_M b} \cdot \nabla\pi \ ,$$

for the equilibrium distribution function, (note that $\beta\Phi$ belongs to the null space of $\overline{\Omega}$), and to

$$\frac{D}{b}\frac{\partial \pi}{\partial t} - \nabla^2 \pi = \Phi \ , \tag{12}$$

for the macroscopic evolution equation associated to the single conservation law existing for this gas.

We find, therefore: (*i*) π satisfies the Poisson equation $\nabla^2\pi = -\Phi$ asymptotically, for $t \gg t_3$. (*ii*) For this to be of any use, the condition $t_3 \ll t_4$ must hold, i.e. $UL \ll b$.

A few remarks follow:

• Eq. (10) can be replaced with Eq. (12). The incompressibility constraint $\nabla \cdot \mathbf{v} = 0$ is built into the model (9,12) for $b \gg L$.

• A queer acoustics remains: As π evolves on a short time scale it forces changes in the N's. In fact,

$$\frac{\partial N_i}{\partial t} \simeq \Omega_i - \gamma \mathbf{c}_i \cdot \nabla \pi$$

yields $\partial \mathbf{v}/\partial t \simeq -\nabla\pi$ when projecting onto the null space of A, while, on the other hand,

$$\frac{D}{b}\frac{\partial \pi}{\partial t} = \nabla^2 \pi + \nabla \cdot \left(\nabla \cdot (\ \mathbf{v}\ \mathbf{v}) - \nabla \frac{v^2}{D} \right) \ .$$

This 'nonlinear mode', however, is naturally suppressed by the numerical procedure, because the N's are frozen while updating π.

• The elliptic character of the problem still underlies the formulations of this and the previous sections. π is not really *determined* by the Poisson equation (or by (12)) since no explicit boundary conditions are available for it. Thus, for the semidiscrete version of (1,2) (discrete time, continuous space),

$$\frac{N_i(\mathbf{x}, t + \Delta t) - N_i(\mathbf{x}, t)}{\Delta t} + \mathbf{c}_i \cdot \nabla N_i(\mathbf{x}, t) = \Omega_i - \gamma \mathbf{c}_i \cdot \nabla \pi$$
$$\nabla \cdot \mathbf{v} = 0 \ ,$$

we would get

$$\nabla^2 \pi = -\Phi(\mathbf{x}, t) + \nabla \cdot \mathbf{v}(\mathbf{x}, t)/\Delta t \ ,$$

but still we must 'guess' the boundary conditions appropriate for the incompressibility constraint to hold at the latter time $t + \Delta t$.

4 Discrete Lattice Gas Model

Now we come to discrete systems. In addition to the primary lattice gas (populations N_i), which accounts for the effects of convection and diffusion, we introduce the secondary gas (populations M_j) generating pressure forces on the primary gas to take care of the incompressibility condition. The M-gas lattice may be a sublattice of the N-gas lattice because its symmetry requirements are weaker. For definiteness, we consider the 24-velocities FCHC lattice for N $(c = \sqrt{2})$. Then a 8-velocities HC sublattice can be used for M (with $b \gg 1$). Time steps for the N and M gases are 1 and $1/b$, respectively. Let $\mathbf{b}_j' = \mathbf{b}_j/b$. For each $\mathbf{c}_i$ there is a unique couple $(\mathbf{b}_{i1}', \mathbf{b}_{i2}')$ such that $\mathbf{c}_i = \mathbf{b}_{i1}' + \mathbf{b}_{i2}'$ and $\mathbf{b}_{i1}' \cdot \mathbf{b}_{i2}' = 0$.

The following discrete versions of the basic differential operators will be used in this section:

$$\nabla\pi(\mathbf{x}) = \frac{D}{b_M}\sum_{i=1}^{b_M}\mathbf{b}_i'\pi(\mathbf{x}+\mathbf{b}_i') = \frac{1}{2}\sum_{i=1}^{b_M}\mathbf{b}_i'\pi(\mathbf{x}+\mathbf{b}_i')$$

$$\nabla\cdot\mathbf{v}(\mathbf{x}) = \frac{1}{2}\sum_{i=1}^{b_M}\mathbf{b}_i'\mathbf{v}(\mathbf{x}+\mathbf{b}_i') \ , \tag{13}$$

where, alternatively, the sums over $\mathbf{b}_i'$ could be changed to sums over $\mathbf{c}_i$. These operators verify the properties:

- $\mathbf{c}_i \cdot \nabla\pi$ belongs to the null space of Ω.
- $\sum_{\text{sites}} \pi\nabla\cdot\mathbf{v} = -\sum_{\text{sites}} \mathbf{v}\cdot\nabla\pi$.

Using these operators we define the discrete analogs of (10) and (9):

$$M_j(\mathbf{x}+\mathbf{b}_j', t+\frac{l+1}{b}) = \frac{1}{8}\sum_k^8 M_k(\mathbf{x}, t+\frac{l}{b}) + \frac{1}{32}\nabla\cdot\boldsymbol{\Psi}(\mathbf{x},t) \tag{14}$$

$$N_i(\mathbf{x}+\mathbf{c}_i, t+1) = N_i(\mathbf{x},t) + A_{ik}\left(N_k(\mathbf{x},t) - N_k^{eq}(\mathbf{x},t)\right)$$
$$-\frac{1}{12}\mathbf{c}_i\cdot\nabla\pi(\mathbf{x},t+1) \ , \tag{15}$$

where $j = 1,\ldots,8$, $i = 1,\ldots,24$ $l = 0,\ldots,b-1$, $\boldsymbol{\Psi} = \mathbf{v}\cdot\nabla\mathbf{v} - \nabla v^2/4 - \mathbf{v}$, and $\pi(\mathbf{x},t+1) = \sum_1^8 M_j(\mathbf{x},t+1)$.

A convenient alternative to the M-gas is the equation

$$\pi(\mathbf{x}, t+2\frac{l+1}{b}) = \frac{1}{8}\sum\pi(\mathbf{x}-2\mathbf{b}_k', t+\frac{2l}{b}) + \frac{\nabla\cdot\boldsymbol{\Psi}}{2} \qquad l = 0,\ldots,b/2-1 \ . \tag{16}$$

The evolution of the whole system is as follows:

- Given the N gas populations at time t, evaluate $\boldsymbol{\Psi}$ and update the M gas (or the equation for π) until a steady state is reached. (Formally b times, but b need not be given in advance). 'Appropriate' conditions at solid boundaries are $\mathbf{n}\cdot\nabla\pi = 0$. (See, eg., [3] for a detailed discussion of this point).
- Advance the N gas one time step using the pressure computed above.

Equation (16) is a Jacobi solver for the Poisson equation. Other iterative schemes conserving locality are available; an obvious candidate would be SOR with odd-even ordering and Chebyshev acceleration, see, eg., [4] . Upon convergence, the scheme yields

$$\nabla^2\pi \equiv \nabla\cdot\nabla\pi = \frac{1}{4}\sum_{i,j=1}^{8}\mathbf{b}_i'\mathbf{b}_j'\pi(\mathbf{x}+\mathbf{b}_i'+\mathbf{b}_j') = -\nabla\cdot\boldsymbol{\Psi} \ . \tag{17}$$

A Chapman-Enskog expansion again shows that the large scale behaviour of this model is described by the Navier-Stokes equations with another two extra terms, in addition to the one already found in Sec. 2, due to the discrete character of the system. (Note however that $-\gamma \sum \mathbf{c}_i(\mathbf{c}_i \cdot \nabla \pi)$ does not give any extra contribution at the Navier-Stokes level). The two new terms are

$$\frac{1}{2}\frac{\partial^2}{\partial t^2}\sum \mathbf{c}_i N_i{}^{eq} = \frac{1}{2}\frac{\partial^2 \mathbf{v}}{\partial t^2} = -\frac{1}{2}\nabla\frac{\partial(\pi - v^2/D)}{\partial t} - \frac{1}{2}\frac{\partial}{\partial t}\nabla\cdot\mathbf{v}\ \mathbf{v}\ ,$$

where the Euler equations have been used to write the last equality, and

$$\frac{1}{2}\sum c_{i\alpha}c_{i\beta}c_{i\gamma}\partial_\beta\partial_\gamma N_i{}^{eq} = \frac{c^2}{2(D+2)}\left(2\nabla(\nabla\cdot\mathbf{v}) + \nabla^2\mathbf{v}\right) = \frac{\nabla^2\mathbf{v}}{6}\ .$$

The second of these terms simply lowers the viscosity to

$$\frac{1}{3}\left(\frac{1}{\lambda} - \frac{1}{2}\right)\ .$$

The other term is less innocuous. It changes the pressure to

$$\left(1 - \frac{1}{2}\frac{\partial}{\partial t}\right)\pi - \frac{v^2}{D}\ ,$$

and changes also the spurious term already present for the continuous case, which becomes

$$\left(\frac{1}{\lambda} + \frac{1}{2}\right)\frac{\partial}{\partial t}\left(\nabla\cdot(\mathbf{v}\ \mathbf{v}) - \frac{\nabla v^2}{D}\right)\ .$$

- For high Reynolds number flows $\delta\pi = O(U^2)$. Hence, each of the new contributions is of the same order as the spurious term found in Sect. 2, (i.e. $O(U^2/LT) = O(U^3/L^2)$).
- For Stokes flows ($G = 0$ in N^{eq}), $\delta\pi = O(U/L)$ and the only spurious term, $(1/2\nabla\partial(\pi/\partial t))$, is too small to appear in the equations anyway.

5 Linear Stability for Stokes Flow (G = 0)

Consider first an infinite medium. Let $-S(\mathbf{x}) = 1/2\sum \overline{N}_i{}^2(\mathbf{x})$, where $\overline{N}_i = N_i - 1/24$. The function $\int -S\, d\Omega$ (for the continuous case of Sect. 2) or $\sum_{\text{sites}} -S(\mathbf{x})$ (for the discrete case of Sect. 4) decreases during the evolution of the system attaining a minimum at equilibrium. For the discrete case the eigenvalues of A must be in $(-2, 0]$.

In fact, for the continuous case:

$$\frac{\partial(-S)}{\partial t} = \sum_i \overline{N}_i\frac{\partial \overline{N}_i}{\partial t} = -\sum_i \overline{N}_i c_{i\alpha}\partial_\alpha \overline{N}_i + \sum_i \overline{N}_i A_{ij}\overline{N}_j - \gamma\sum_i \overline{N}_i c_{i\alpha}\partial_\alpha \pi\ ,$$

where

$$-\sum_i \overline{N}_i c_{i\alpha} \partial_\alpha \overline{N}_i = -\frac{1}{2} \partial_\alpha \sum_i c_{i\alpha} \overline{N}_i^{\,2}$$

vanishes on integrating,

$$\sum_i \overline{N}_i A_{ij} \overline{N}_j \leq 0 \ ,$$

the equality holding only for $\overline{N}_i = \overline{N}_i^{\,eq} = \mathbf{c}_i \cdot \mathbf{v}/12$, and

$$-\gamma \sum_i \overline{N}_i c_{i\alpha} \partial_\alpha \pi = -\gamma \partial_\alpha \left(\sum_i c_{i\alpha} \overline{N}_i \pi \right) + \gamma \pi \sum_i c_{i\alpha} \partial_\alpha \overline{N}_i \ .$$

Here the first term on the right side vanishes on integrating, and the second term is simply $\gamma \pi \nabla \cdot \mathbf{v} = 0$.

For the discrete case:

$$\sum_{\text{sites}, i} \overline{N}_i^{\,2}(\mathbf{x}, t+1) = \sum_{\text{sites}, i} \overline{N}_i^{\,2}(\mathbf{x} + \mathbf{c}_i, t+1) =$$

$$\sum_{\text{sites}, i} \left\{ \overline{N}_i(\mathbf{x}, t) + A_{ij} \overline{N}_j(\mathbf{x}, t) - \frac{(\mathbf{c}_i \cdot \nabla \pi)}{12} \right\}^2 =$$

$$\sum_{\text{sites}} \left\{ \sum_i \overline{N}_i^{\,2} + 2 \sum_{ij} \overline{N}_i A_{ij} \overline{N}_j + \sum_{ijk} A_{ij} \overline{N}_j A_{ik} \overline{N}_k \right\}$$

$$+ \sum_{\text{sites}} \left\{ -\frac{1}{6} \left(\sum_i \overline{N}_i \mathbf{c}_i \cdot \nabla \pi + \sum_{ij} A_{ij} \overline{N}_j \mathbf{c}_i \cdot \nabla \pi \right) + \frac{\sum (\mathbf{c}_i \cdot \nabla \pi)^2}{12^2} \right\}$$

$$= \sum_{\text{sites}} \left\{ \sum_\mu (1 + \lambda_\mu)^2 \overline{N}_\mu^{\,2} - \frac{1}{6} \mathbf{v} \cdot \nabla \pi + \frac{|\nabla \pi|^2}{12} \right\}$$

$$= \sum_{\text{sites}} \left\{ \sum_\mu (1 + \lambda_\mu)^2 \overline{N}_\mu^{\,2} + \frac{\pi}{6} (\nabla \cdot \mathbf{v} - \frac{\nabla^2 \pi}{2}) \right\}$$

$$= \sum_{\text{sites}} \left\{ \sum_\mu (1 + \lambda_\mu)^2 \overline{N}_\mu^{\,2} - \frac{|\nabla \pi|^2}{12} \right\} \ ,$$

where λ_μ are the eigenvalues of A and $\overline{N}_\mu$ are the projections of $\overline{N}$ along the corresponding eigenvectors. Use has been made of (17) and of the properties of the discrete ∇. As can be seen, $\sum \overline{N}_i^{\,2}(t+1) \leq \sum \overline{N}_i^{\,2}(t)$ if $-2 < \lambda_\mu \leq 0$, the equality holding only at equilibrium.

Non-slip boundary conditions at rigid walls are known to introduce spurious numerical boundary layers for many classical numerical schemes for the solution of the incompressible Navier-Stokes equations. These layers are analogous to the

Knudsen layer for a lattice gas, or for a gas with discrete velocities. The effect of solid boundaries can be analysed in the framework of the two-dimensional incompressible flow in a plane channel, $-L < x < L$ with $L \gg 1$, obtained by considering a solution of the linear equations periodic along the channel, of the form

$$(N_i,\, \pi)(x, y, t) = \{N_i(x,t),\, \pi(x,t)\} \exp(iky) \ , \tag{18}$$

for some real wavenumber k (with $kL = O(1)$). Here $x = \pm L$ are the solid boundaries (see, e.g. Orszag, Israeli and Deville [3]).

The essential features already appear to a sufficient extent in a semidiscrete model. To simplify matters further, a rectangular HPP lattice is considered instead of the 2D projected FCHC.

Writing, for the completely continuous case,

$$\begin{aligned} &(N_i,\, \pi) = \{N_i(x),\, \pi(x)\} \exp(ikx + \sigma t), \quad i = 1, \ldots, 4 \\ &N_1 = N_3, \quad N_2 = N_4 \text{ at } x = \pm L \ , \end{aligned} \tag{19}$$

we get

$$\begin{aligned} \sigma N_1 + \tfrac{1}{\sqrt{2}}\left(\tfrac{dN_1}{dx} + ik\, N_1\right) &= \Omega_1 - \frac{1}{2\sqrt{2}}\left(\frac{d\pi}{dx} + ik\, \pi\right) \\ \sigma N_2 + \tfrac{1}{\sqrt{2}}\left(-\tfrac{dN_2}{dx} + ik\, N_2\right) &= \Omega_2 - \frac{1}{2\sqrt{2}}\left(-\frac{d\pi}{dx} + ik\, \pi\right) \\ \sigma N_3 + \tfrac{1}{\sqrt{2}}\left(-\tfrac{dN_3}{dx} - ik\, N_3\right) &= \Omega_3 - \frac{1}{2\sqrt{2}}\left(-\frac{d\pi}{dx} - ik\, \pi\right) \\ \sigma N_4 + \tfrac{1}{\sqrt{2}}\left(\tfrac{dN_4}{dx} - ik\, N_4\right) &= \Omega_4 - \frac{1}{2\sqrt{2}}\left(\frac{d\pi}{dx} - ik\, \pi\right) \ . \end{aligned} \tag{20}$$

The matrix of the linearised collision operator is

$$\Omega = -\frac{1}{4} \operatorname{circ}\,(\alpha + \beta,\, \alpha - \beta,\, \alpha + \beta,\, \alpha - \beta) \ . \tag{21}$$

where $-2 < (\alpha,\, \beta) < 0$ are the nonzero eigenvalues associated to $(1,1,1,1)$ and $(1,-1,1,-1)$, respectively.

Solutions of the form $\exp(\lambda x)$ exist with λ given by a biquadratic equation. Writing $(\lambda_+ L)^2 = a + b$, $(\lambda_- L)^2 = a - b$, the dispersion relation becomes

$$\frac{\tanh \sqrt{a+b}}{\sqrt{a+b}} = \frac{\tanh \sqrt{a-b}}{\sqrt{a-b}} \ ,$$

so that $(a, b) = O(1)$ and the two possible values of the growth rate are $\sigma = -\beta + \epsilon$ or $\sigma = -\epsilon$, with $0 < \epsilon = O(1/L^2) \ll 1$.

No Knudsen layer appears in this case, since $\lambda L = O(1)$. This is a consequence of the extreme simplicity of the HPP model; for the FHP or FCHC models, for example, a Knudsen layer does exist. The author is indebted to D. d'Humieres for pointing out to him that this should indeed be the case.

Let us turn now to the semidiscrete model. The computation of the pressure can be understood as a time splitting procedure, in which the evolution obeys the two successive steps

$$\begin{aligned} &N_i{}^*(\mathbf{x}) - N_i(\mathbf{x},t) = -\gamma \mathbf{c}_i \cdot \nabla\pi \\ &\mathbf{v}^* = \sum \mathbf{c}_j N_j{}^*, \quad \nabla\cdot\mathbf{v}^* = 0, \quad \mathbf{v}^* = 0 \text{ at } x = \pm L \ , \end{aligned} \tag{22}$$

and

$$N_i(\mathbf{x}, t+1) - N_i{}^*(\mathbf{x}) + \mathbf{c}_i \cdot \nabla N_i = \Omega_i \ . \tag{23}$$

Schemes of this type for the Navier-Stokes equations are prone to introduce numerical boundary layers responsible for substantial time difference errors. In fact, looking for a solution of the form

$$(N_i,\, N_i{}^*,\, \pi)(x,y,n) = \{N_i(x),\, N_i{}^*(x),\, \pi(x)\}\ \exp(iky)K^n \ , \tag{24}$$

with $\{N_i(x),\, N_i{}^*(x),\, \pi(x)\} \sim \exp(\lambda x)$, one is left with a characteristic equation cubic in λ^2. Two of its solutions are of the order of $1/L^2$, but the other is of the order of the unity and, therefore, leads generically to numerical boundary layers near the solid walls.

References

1. Frisch, U., d'Humieres, D., Hasslacher, B., Lallemand, P., Pomeau, Y., and Rivet, J.P.: Lattice gas hydrodynamics in two and three dimensions. Complex Systems **1** (1987) 646
2. Higuera, F.J. and Jiménez, J.: Boltzmann approach to lattice gas simulations. Europhys. Lett. **9** (1989) 663
3. Orszag, S., Israeli, M., Deville, M.: Boundary conditions for incompressible flows. J. Sci. Comput. **1** (1986) 75
4. Press, W., Flannery, B., Teukolsky, S., and Vetterling, W.: *Numerical Recipes*. Cambridge University Press 1986

This article was processed using the LaTeX macro package with ICM style

Obtaining Numerical Results from the 3D FCHC-Lattice Gas

J.A. Somers and P.C. Rem

Koninklijke/Shell-Laboratorium, Amsterdam (Shell Research B.V.)
Badhuisweg 3, 1031 CM Amsterdam, The Netherlands.

Submitted to the proceedings of the Shell 'Workshop on Numerical Methods for the Simulation of Multi-Phase and Complex Flow', to appear in Springer Proceedings on Physics, Ed. T.M.M. Verheggen.

1 Introduction

In 1986 the hexagonal lattice gas model was introduced as a new technique to simulate fluid flow phenomena satisfying the two-dimensional incompressible Navier-Stokes equations [1]. Soon afterwards, a three-dimensional variant based on the FCHC lattice was introduced [2], starting off several initiatives to try and solve realistic flow problems.

In the past few years quite some knowledge has been obtained about the algorithmic and numerical properties of the lattice gas scheme. Still the evolution of lattice gas models has not finished yet. Regularly, new improvements are suggested, both in the context of numerical effectiveness and concerning the algorithmic efficiency. Nevertheless, we have now reached a state, where we can try to assess the relative value of the lattice gas technique as compared to the potential of other numerical schemes. In this paper we want to address some of the numerical and algorithmic properties of the FCHC-lattice gas scheme, in order to identify its scope of application.

From a numerical point of view the lattice gas technique is very similar to a space- and time-explicit finite difference scheme [3, 4]. In [5] it was argued that the order of the scheme with respect to the accuracy of the solution, would be only 1.5. Yet, one should realize that the leading cause of inaccuracy in the solution of a lattice gas simulation is noise. Noise is strongly present in the microscopic representation of the solution, but does not affect fundamentally the equations that are being solved, nor does it deteriorate the convergence or numerical stability of the scheme. Thus, even a less accurate solution of a lattice gas simulation may still suit its purpose. Another important issue in the application of any space- and time-explicit finite difference scheme to a particular flow problem is the scaling of the features. The lattice gas technique is not capable of adjusting its resolution to local features of the flow, nor can it solve fast features implicitly. Alternative techniques may improve on this, though often at the cost of other compromises (such as the proper handling of boundary conditions).

Considering the implementation of the scheme it should be noted that most computation time is spent in the discrete state transitions of the nodes in the lattice. In principle such a state transition can be implemented very efficiently via a table lookup, albeit that the state space of the FCHC lattice involves 16777216 entries and hence an enormous amount of memory is needed to store such a state transition table. Many authors have already suggested algorithms that reduce the requirements of overhead storage, at the cost of computational efficiency or non-optimal physical properties of the state transition rules [6, 7, 8, 9]. In this series of improvements we present a new algorithm which performs the state transition of the FCHC-lattice gas without introducing any conflicts with the design criteria of the lookup table. The overhead storage is limited ($<$ 0.5Mbytes) and the loss of computational efficiency is marginal.

In the second section we present the new FCHC-algorithm, which is capable of running any of the optimal semi-detailed-balance models [10] and the models without semi-detailed-balance [7, 11] on an ordinary workstation. In the third section we give a summary of the results that have been obtained so far in comparing the behaviour of our model with the predictions of kinetic theory. Besides some relevant technical details these two sections illustrate that one cannot make a final judgement yet about the usefulness of the lattice gas technique, as there is still room for a better understanding. The fourth section presents an example of how the lattice gas can be used as a numerical tool to solve a realistic flow problem; in this section we will discuss its strengths and weaknesses, and derive a scope of its applicability. Finally, in the fifth section, we summarize the conclusions.

2 A new algorithm for the FCHC-lattice gas

The FCHC-lattice gas automaton tracks the evolution of a microscopic world of particles, moving synchronously at unit speed along the edges of the FCHC lattice [2, 12]. Originally the FCHC lattice is a four-dimensional lattice, but for the lattice gas application the fourth dimension has been curved microscopically into two layers with periodic boundary conditions, and only three dimensions are expanded, spanning the three-dimensional space (see Figure 1). Each node in this lattice is connected to 24 neighbouring nodes, which are reached along the velocity directions $\mathbf{c}_i$. We have chosen the following ordering of the $\mathbf{c}_i$, suiting our later purposes.

$$
\begin{array}{lll}
\mathbf{c}_0 = (+1,+1,0,0) & \mathbf{c}_4 = (+1,0,+1,0) & \mathbf{c}_8 = (+1,0,0,+1) \\
\mathbf{c}_1 = (+1,-1,0,0) & \mathbf{c}_5 = (+1,0,-1,0) & \mathbf{c}_9 = (+1,0,0,-1) \\
\mathbf{c}_2 = (0,0,+1,+1) & \mathbf{c}_6 = (0,+1,0,+1) & \mathbf{c}_{10} = (0,+1,+1,0) \\
\mathbf{c}_3 = (0,0,+1,-1) & \mathbf{c}_7 = (0,+1,0,-1) & \mathbf{c}_{11} = (0,+1,-1,0)
\end{array}
$$

$$
\begin{array}{lll}
\mathbf{c}_{12} = (-1,-1,0,0) & \mathbf{c}_{16} = (-1,0,-1,0) & \mathbf{c}_{20} = (-1,0,0,-1) \\
\mathbf{c}_{13} = (-1,+1,0,0) & \mathbf{c}_{17} = (-1,0,+1,0) & \mathbf{c}_{21} = (-1,0,0,+1) \\
\mathbf{c}_{14} = (0,0,-1,-1) & \mathbf{c}_{18} = (0,-1,0,-1) & \mathbf{c}_{22} = (0,-1,-1,0) \\
\mathbf{c}_{15} = (0,0,-1,+1) & \mathbf{c}_{19} = (0,-1,0,+1) & \mathbf{c}_{23} = (0,-1,+1,0)
\end{array}
$$

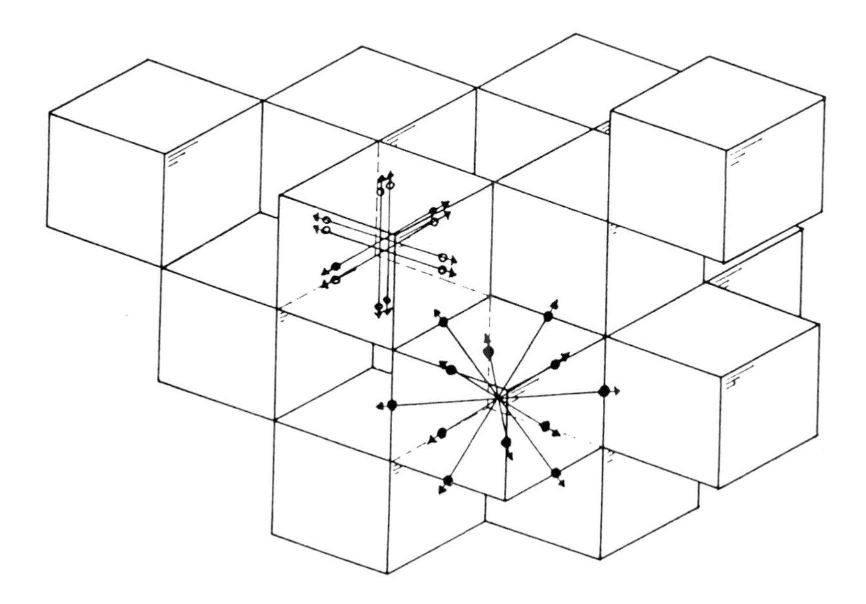

Fig. 1. The structure of the FCHC lattice with a two-layer periodicity in the fourth dimension. The nodes of the lattice are the centers of cubic cells. The two three-dimensional layers are interleaved, such that each node has 12 connections with 12 different neighbouring nodes in its own layer, and 12 connections with only 6 different nodes in the other layer. The figure shows only one layer, i.e. the 'empty' three-dimensional space is filled by the other layer.

During the evolution of the automaton, particles gather at each node in the lattice from the 24 neighbouring nodes along $-\mathbf{c}_i$ and exchange momentum in a collision. In the next step they propagate towards the neighbouring node along $\mathbf{c}_i$ and collide again with other particles. The sequence of one *propagation step* and one *collision step* is identified as a *time step* in the evolution of the lattice gas.

The exchange of momentum during the collision is prescribed by the *state transition rules*. The presence or absence of each of the 24 gathering particles is encoded with a single bit. So, the whole set of possible particle configurations prior to the collision contains 2^{24} elements. The state transition rules should specify an after-collision configuration for each of the possible pre-collision configurations. One solution is to provide a huge table consisting of 2^{24} after-collision states, which can be accessed using the integer representation of the pre-collision state [13].

This solution is not feasible with computers that do not have the required amount of memory available. However, there is a similarity within the state transition rules that can be exploited in order to actually use a much smaller table of after-collision states. The similarity originates from the symmetry properties of the lattice. It has been shown that the optimization criteria for choosing a particular after-collision state for a certain pre-collision state are invariant with respect to all lattice isometries [14]. In fact, this means that if the transition rules specify that s^0 should collide into s^1, then, for any lattice isometry p, the state $p(s^0)$ may collide into the state $p(s^1)$, without any loss of generality.

The FCHC lattice exhibits 1152 different isometries, and if we use these isometries to eliminate all equivalents from the 16 million possible particle configurations, only 18736 essentially different states remain. So in fact we only have to specify the transition rules for these 18736 states. During the evolution of the lattice gas each of the pre-collision states should be permuted first by an appropriate isometry, before one of the 18736 transition rules can be applied. Finally, the result of this transition should be permuted by the inverse isometry yielding the after-collision state.

However, there is no efficient algorithm available to decide which of the 1152 lattice isometries should be chosen for any particular case. In the following we present an algorithm, which selects with little computation an isometry such that the permuted pre-collision state will reside in a set of 106496.

Consider the standard way of representing a state with a 24-bit integer s. The ith bit in this integer indicates whether a particle with velocity $\mathbf{c}_i$ is present or not. Lattice isometries can be applied to such a representation just by permuting the bits in the integer. It has been observed that all lattice isometries leave invariant the geometrical angles between any two velocity directions $\mathbf{c}_i$ and $\mathbf{c}_j$ [14] and that, in particular, particles on $\mathbf{c}_i$ and $-\mathbf{c}_i$ stick together when such a permutation is applied. This observation has inspired us to choose a different representation, which explicitly relates the configuration of particles on 12 $\mathbf{c}_i$-s to that of their $-\mathbf{c}_i$-counterparts.

We represent a 24-particle configuration uniquely with two 12-bit integers s_d and s_s. The ith bit in s_d indicates whether the occupancy of edge $\mathbf{c}_i$ differs from the occupancy along $-\mathbf{c}_i$, i.e. whether a particle is present on exactly one of the edges $\mathbf{c}_i$ and $-\mathbf{c}_i$. The ith bit in s_s specifies whether a particle is present with velocity $\mathbf{c}_i$, just as in the standard representation. Early in this section we have already ordered the $\mathbf{c}_i$, such that $\mathbf{c}_{i+12} = -\mathbf{c}_i$ for all $i : 0 \leq i < 12$. In this way we can easily translate states between the standard representation

$s : 0 \leq s < 2^{24}$ and the aligned representation $(s_d, s_s) : 0 \leq s^d, s^s < 2^{12}$, which we have introduced above.

$$s_d = (s \text{ div } 4096) \text{ xor } (s \text{ mod } 4096) \qquad (1)$$
$$s_s = s \text{ mod } 4096$$
$$s = 4096 * (s_d \text{ xor } s_s) + s_s$$

The effect of a lattice isometry on a state in the aligned representation is slightly more complicated than just a permutation on bits. We already indicated that particles on $\mathbf{c}_i$ and $-\mathbf{c}_i$ remain together during the permutation, thus the permutation on the s_d part of a state will be identical to the permutation on the s_s part. Hence, the isometry is represented by a permutation on 12-bit integers, to start with. Any isometry leaves the number of particles moving along $\mathbf{c}_i$ or $-\mathbf{c}_i$ invariant, so the value of s_d is not affected otherwise. However, the isometry might swap the particles on the $\mathbf{c}_i$ side and the $-\mathbf{c}_i$ side, so if precisely one of those edges carries a particle, the value s_s may need to be changed. In practice we represent each lattice isometry p by a pair (p_p, p_f), in which p_p specifies a permutation of the bits in a 12-bit integer and p_f is a 12-bit integer indicating which of the $(\mathbf{c}_i, -\mathbf{c}_i)$ pairs will have swapped its particles, prior to the permutation. The effect of an isometry p and its inverse p^{-1} on a state (s_d, s_s) is defined by

$$p(s_d, s_s) = (\ p_p(s_d)\ ,\ p_p(s_s \text{ xor } (p_f \text{ and } s_d))\ ) \qquad (2)$$
$$p^{-1}(s_d, s_s) = (\ p_p^{-1}(s_d)\ ,\ p_p^{-1}(s_s) \text{ xor } (p_f \text{ and } p_p^{-1}(s_d))\ )$$

Now, notice that the effect of any isometry p on the s_d part of a pre-collision state only depends on that very same 12-bit s_d part! The question arises: "How many of the possible 4096 s_d parts in the pre-collision state are essentially different, when we eliminate all equivalents with respect to the 1152 isometries?" The answer is that only 51 representative s_d parts remain. So we can easily reduce the 4096×4096 state space down to a 51×4096 state space. A priori we construct two tables `tabP[0..4095]` and `tabT[0..4095]` storing for each possible value of s_d an isometry that maps s_d onto one of the 51 representatives and a sequence number which identifies the representative that it will be mapped onto.

The next question is: "Are all 1152 isometries equally relevant in reducing the s_d state space from 4096 to 51?" The answer is negative. The largest equivalence class contains only 288 s_d parts, which is the case for four of the 51 representatives. One isometry that is not needed in any of the isometries in `tabP` is the reversal of all coordinates, r_{xyzt}.

$$x \mapsto -x \quad y \mapsto -y \qquad (3)$$
$$z \mapsto -z \quad t \mapsto -t$$

This isometry swaps all particles on each $\mathbf{c}_i$ and $-\mathbf{c}_i$, so it transforms a state in the following way.

$$r_{xyzt}(s_d, s_s) = (s_d\ ,\ s_s \text{ xor } s_d) \qquad (4)$$

Indeed, the s_d part of a state is not affected by this isometry, thus it is of no use in the reduction of the s_d state space.

The r_{xyzt} can be used, however, to reduce the state space of the s_s part of the pre-collision states. If we can guarantee that each of the 51 s_d representatives has a value of at least 2048 (i.e. the most significant bit of the integer representation is 1), then we can toggle the most significant bit of s_s by applying the r_{xyzt} isometry after the isometry from `tabP`. In that case we can reduce the state space of the s_s part from 4096 to 2048.

From each of the 51 s_d equivalence classes the maximum integer value is selected as its representative. Fifty of them will have a value of at least 2048. So we can now code an algorithm that requires a state transition table for a state space of size 52×2048 and uses the lattice isometries in order to derive a sequence number between 0 and 50 and a reduced s_s part from any of the 2^{24} possible pre-collision states. Before giving the complete collision algorithm, we will first specify all the tables that should be constructed a priori.

`permP[0..1151]`	A lattice isometry is fully specified by a bit-permutation on 12-bit integers, and a flip indicator for the s_s part of a state.
`permF[0..1151]`	This table contains the flip indicators of each of the 1152 lattice isometries.
`permI[0..1151]`	This table contains for each lattice isometry the pre-computed index of its inverse.
`tabR[0..50]`	This table contains the 51 representatives of the s_d equivalence classes. The first 50 representatives all have their most significant bit non-zero. The 51st representative is zero.
`tabC[0..51][0..2047]`	The after-collision states are stored in 52 tables of 2048 integers each. As only the 51st representative cannot reduce its s_s part to below 2048, all 4096 after-collision states are provided in `tabC[50]` and `tabC[51]`.
`tabP[0..4095]`	The isometry that transforms a 12-bit s_d part into one of the 51 representatives is pre-computed in this table. This isometry will not involve the r_{xyzt} reversal.
`tabT[0..4095]`	This table stores for each s_d the sequence number of the representative that the isometry in `tabP` will result in.

The collision algorithm starts from the pre-collision state `spre` in the 24-bit standard representation. We will apply a random pre-permutation `prep`, in order to ensure microscopic isotropy of the collision operator. The permutation on 12-bit integers is performed by an integer function `permute12(s,pnr)`.

```
function collide(spre,prep: integer): integer;
var sd,ss,tnr,pnr: integer;
     tnr,pnr,ipnr,iprep: integer;
begin
    ss:=spre mod 4096
   ;sd:=(spre div 4096) xor ss
   ;ss:=permute12(ss xor (sd and permF[prep]), prep)
   ;sd:=permute12(sd, prep)
   ;tnr:=tabT[sd]; pnr:=tabP[sd]
   ;ss:=permute12(ss xor (sd and permF[pnr]), pnr)
   ;if ss<2048
      then begin sd:=tabC[tnr][ss] div 4096
      ;ss:=tabC[tnr][ss] mod 4096 end
      else begin ss:=ss xor tabR[tnr]
      ;sd:=tabC[tnr][ss] div 4096; ss:=tabC[tnr][ss] mod 4096
      ;ss:=ss xor sd end
   ;ipnr:=permI[pnr]
   ;ss:=permute12(ss xor (sd and permF[ipnr]), ipnr)
   ;sd:=permute12(sd, ipnr)
   ;iprep:=permI[prep]
   ;ss:=permute12(ss xor (sd and permF[iprep]), iprep)
   ;sd:=permute12(sd, iprep)
   ;collide:=4096*sd+ss
end;
```

On a Sun Sparc station 1 this algorithm runs at 40000 node updates per second. Most of our simulations are run on the Parsytec FT400 parallel computer, based on 400 transputers. This machine is capable of sustaining 2.10^6 node updates per second. Some effort is required in rewriting this algorithm such that high performances can be achieved on a vector computer.

In this section, we did not discuss how to compute the state transition rules for the table tabC. This is not a trivial exercise, though. Details can be found in [6, 7, 10, 11, 14, 15].

3 Numerical measurement of the transport coefficients

Dubrulle et al. have shown that the transport coefficients of the lattice gas models without semi-detailed balance do not match the theoretically predicted values [11]. Therefore we will first present some results that we have obtained from measuring the effective transport coefficients of our FCHC transition rules introduced in [7]. Indeed we have been capable of reproducing their observations. Furthermore, we present some early results that indicate how the Boltzmann approximation can be recovered partly by artificially removing correlations.

Two transport coefficients, $g(\rho)$ and ν, appear in the macroscopic flow equations for the FCHC-lattice gas evolution in the incompressible limit.

$$\rho \partial_t \mathbf{u} + g(\rho) \rho \mathbf{u} \nabla \mathbf{u} = -\nabla p + \rho \nu \Delta \mathbf{u} \tag{5}$$

The $g(\rho)$ coefficient is an artefact of the lattice gas models and will prove irrelevant after scaling the velocity. The ν-coefficient is the kinematic viscosity. It is important to know the effective values for both transport coefficients, as they play a key role in scaling the lattice gas variables to real physical quantities.

First we will determine the $g(\rho)$ transport coefficient from the equilibrium distribution of particles N_i of all 24 velocity directions $\mathbf{c}_i$. Therefore we have run a uniform flow on a lattice with size $30 \times 30 \times 30$, with periodic boundary conditions in all directions. The mean particle distribution has been averaged over the whole lattice during 5000 time steps. The kinetic theory of lattice gases predicts how the N_i should be related to the density, velocity and $g(\rho)$ [15]. However, the N_i-values that we have measured in our simulation definitely do not fit this functional form. Two issues play a role here. First, the theory is based on the Boltzmann approximation, which does not take into account particle correlations. Secondly, the microscopic periodicity of the lattice in the fourth dimension introduces a spatial anisotropy in these particle correlations.

The correlations can effectively be removed from the simulation by simultaneously running multiple lattices in an ensemble and incorporating a well controlled mixing of particles among the lattices during the propagation step. An ensemble with two lattices can remove all correlations after one time step due to the periodicity of the fourth dimension. Hénon has shown that thereafter precisely 24 lattices are needed to remove all correlations after two time steps [16]. The explicit combination of multiple lattices in an ensemble-like system is not new. In [17] it has already been reported that ensemble simulations of two-phase lattice gas models show interfaces with little noise.

We have studied ensembles with one, two and four lattices, eliminating all correlations after one time step and a (isotropic) subset of the correlations after two time steps. The correlations that have not been removed still induce an asymmetry in the $g(\rho)$-factor. This asymmetry can be captured from the measurements by separately fitting the N_i values of particles that have a velocity component in the fourth dimension and those that have not, yielding respectively $g^4(\rho)$ and $g(\rho)$. The following table enumerates these fitted values for the three ensemble sizes separately. Remember that these values do depend on the precise rules that we have used for the state transitions.

	theory $g(\rho)$	*ensemble 1* $g(\rho)/g^4(\rho)$	*ensemble 2* $g(\rho)/g^4(\rho)$	*ensemble 4* $g(\rho)/g^4(\rho)$
$\rho = 7.0$	0.449	0.442/0.390		
$\rho = 8.0$	0.410	0.387/0.325	0.393/0.375	0.397/0.389
$\rho = 9.0$	0.366	0.332/0.278	0.338/0.300	0.345/0.321
$\rho = 10.0$	0.317	0.268/0.220	0.276/0.241	0.288/0.271
$\rho = 11.0$	0.259	0.195/0.161	0.207/0.179	0.221/0.214
$\rho = 12.0$	0.191	0.114/0.109	0.130/0.124	0.146/0.140
$\rho = 13.0$	0.109	0.024/0.051	0.043/0.056	0.056/0.065
$\rho = 14.0$	0.009	-0.079/-0.022	-0.060/-0.023	

Notice that the asymmetry is less apparent for the largest ensemble size in this table. Also at a density near 12 isotropy seems to be recovered. The measured values for $g(\rho)$ do not compare very well with the predicted theoretical value. However, as the ensemble size increases, the improvement is significant. This can be observed more clearly from the graph in Figure 2.

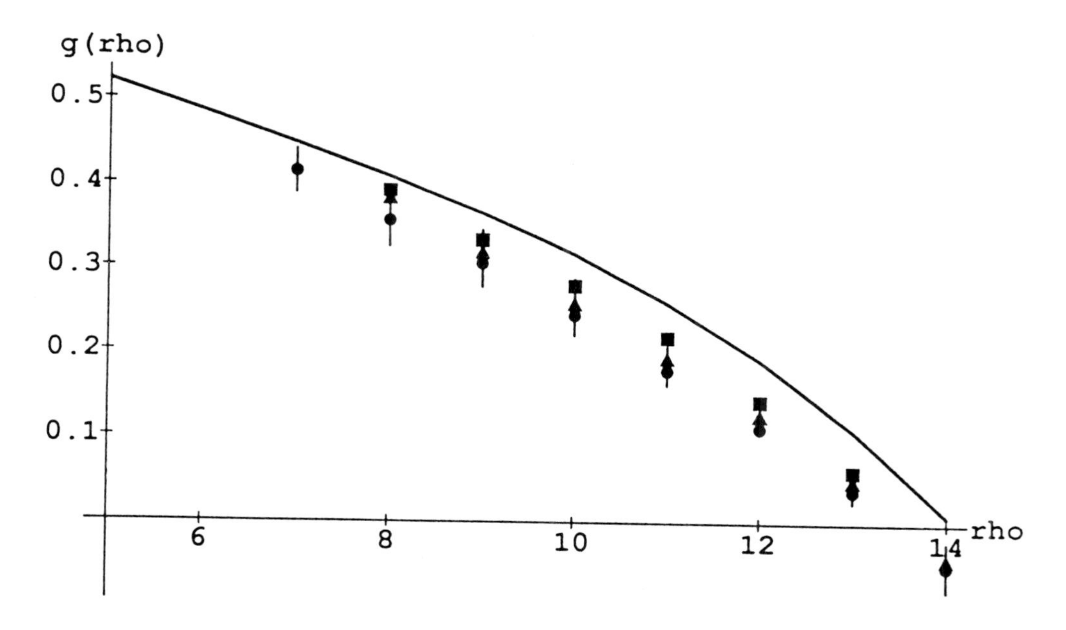

Fig. 2. The $g(\rho)$ transport coefficient for our set of FCHC state-transition rules. The dots, triangles and squares indicate the values obtained from simulations with respectively one, two and four lattices in the ensemble. The solid line indicates the prediction, based on the Boltzmann approximation.

Next we will present our results in obtaining the kinematic viscosity ν experimentally. We have simulated a standard Poiseuille flow between two solid walls on a lattice of size $60 \times 20 \times 20$. Periodic boundary conditions apply in two dimensions, parallel to the solid walls. The distance between the solid boundaries is 18 lattice units.

The flow is driven by a force which is uniformly distributed over the whole volume. The effect of this force is macroscopically equivalent with a linear pressure drop in the flow. The force is implemented by incidentally flipping two bits in the particle configuration after the collision, such that momentum in a certain direction is added to the microscopic state of a node in the lattice. Let f denote the norm of the average momentum added per node in the lattice per time step by this bit flipping. In the case of our steady-state Poiseuille flow an explicit relation between the viscosity ν, the applied force f, the mean velocity u, and the channel diameter D can be derived.

$$\nu = \frac{D^2 f}{12 \rho u} \tag{6}$$

Again we are interested in the effect of the ensemble size on the viscosity. In the following table we summarize the results.

	theory	*ensemble 1*	*ensemble 2*	*ensemble 4*
	ν	ν	ν	ν
$\rho = 8.0$	0.01563	0.0229	0.0201	0.0195
$\rho = 9.0$	0.01031	0.0192	0.0163	0.0150
$\rho = 10.0$	0.00666	0.0142	0.0127	0.0108
$\rho = 11.0$	0.00451	0.0121	0.0101	0.0087
$\rho = 12.0$	0.00380	0.0105	0.0090	0.0079
$\rho = 13.0$	0.00451	0.0106	0.0093	0.0078

These values have been plotted in the graph of Figure 3. The consequences of the deviations of the actual transport coefficients $g(\rho)$ and ν from their predicted theoretical values can best be appreciated from values of the resulting R_* coefficient, which are plotted in Figure 4. The optimal R_* value of about 42 is not reached by far. Instead we have observed maxima of 13.3, 15.9 and 18.9 for respectively one, two and four lattices in the ensemble. (The R_*-coefficient expresses the 'quality' of the lattice gas model in the sense that a flow at Mach number M with typical length scale L will show a Reynolds number $Re = R_* M L$.)

4 The interpretation of an example application

In this section we will give an idea of the kind of complexity that can be tackled with the three-dimensional lattice gas technique. An important aspect of applying the lattice gas automaton results to a realistic flow experiment is the scaling of the dimensionless variables to real physical quantities. This scaling exercise more or less reveals the scope of application.

First let us describe the example flow problem of this section. The white surface in Figure 5 depicts a three-dimensional geometry, which is bounded at

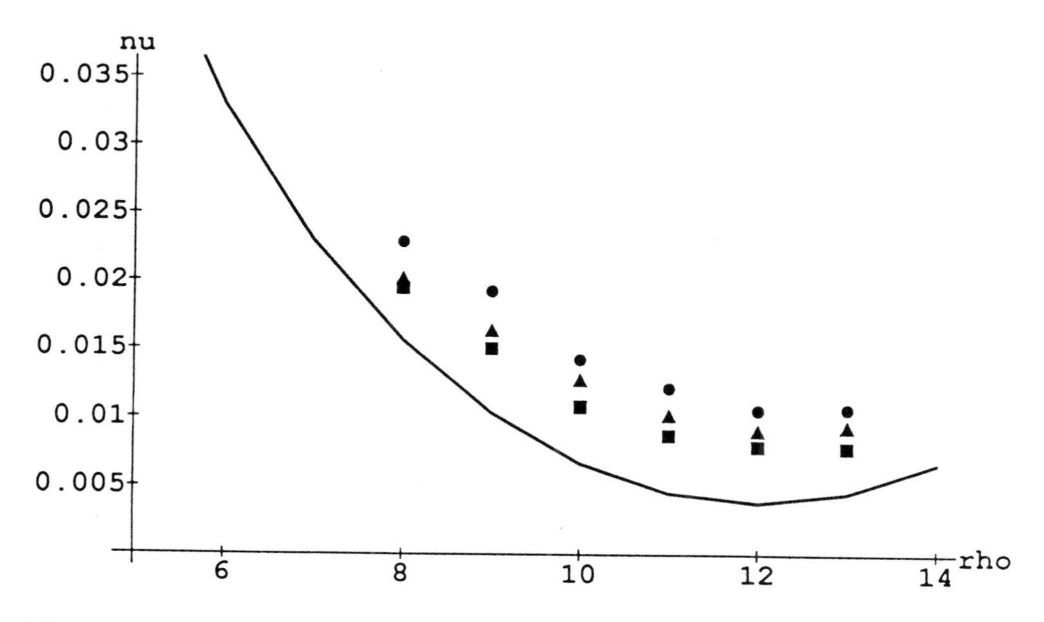

Fig. 3. The kinematic viscosity ν for our set of FCHC state transition rules. The dots, triangles and squares depict the measured values obtained from simulations with respectively one, two and four lattices in the ensemble. The statistical error in the measurements was less than the size of the markers. The solid line indicates the prediction based on the Boltzmann approximation.

the top and from below by two plates with stick boundary conditions. A feed (say water) is pushed through horizontally at a 45° angle to the x-axis. Periodic boundary conditions apply both in the x- and in the y-direction. The dimensions of the box are $4.4mm \times 4.4mm \times 1.0mm$. The superficial velocity is $0.31m/s$ establishing a Reynolds number of 310, based on the height of the box. We will drive the flow with a body force.

Our job is to compute the flow between the horizontal planes explicitly, i.e. in a first stage we want to compute the complete time-dependent velocity profile and the pressure field around the geometry. But later we want to extract other results from these data, such as the vorticity, the distribution of the turbulent energy as a function of the distance to the horizontal plates, the spectrum of turbulent scales, the drag of the geometry, etc.

In order to assess whether this problem can be solved with the lattice gas technique one first has to scale the problem specification into the lattice gas phase space. Basically, there are six scaling relations (see also [12]).

$$
\begin{aligned}
t &= \tau t^{\mathrm{ca}} & p &= p_0 + \alpha_p p^{\mathrm{ca}} \\
L &= L^{\mathrm{ca}} & \mathrm{u} &= \alpha_u \mathrm{u}^{\mathrm{ca}} \\
\rho &= \alpha_\rho \rho^{\mathrm{ca}} & \nu &= \alpha_\nu \nu^{\mathrm{ca}}
\end{aligned}
\tag{7}
$$

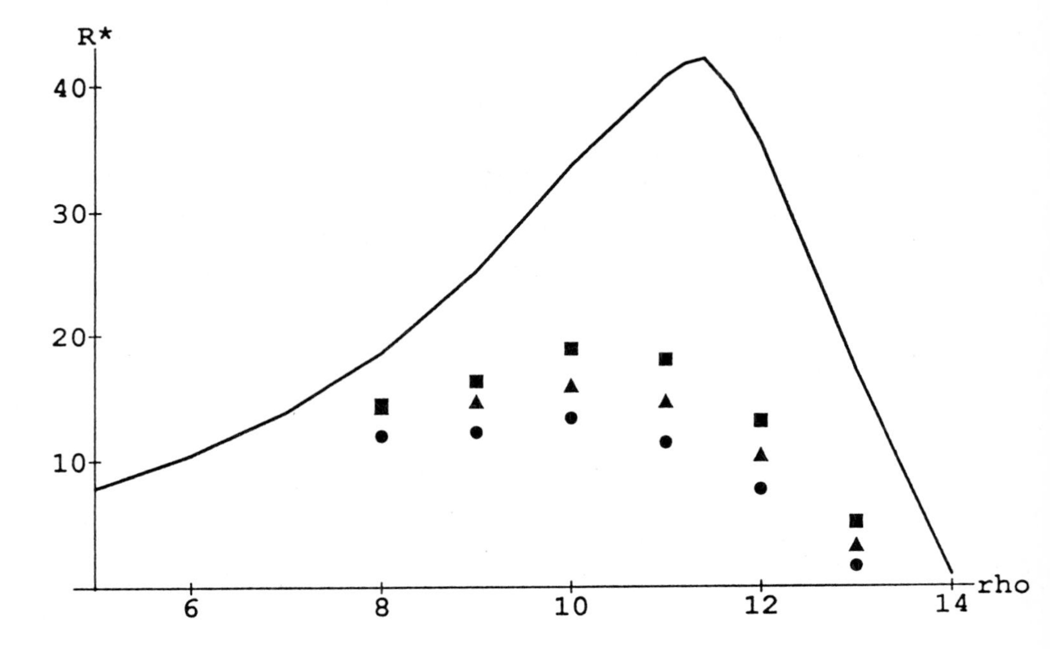

Fig. 4. Effective Reynolds coefficient R_* for our version of the FCHC model. The dots, triangles and squares depict the values obtained from simulations with respectively one,two and four lattices in the ensemble. The solid line indicates the prediction based on the Boltzmann approximation.

These relations map the macroscopic lattice gas automaton quantities time t^{ca}, space L^{ca}, density ρ^{ca}, pressure p^{ca}, velocity $\mathbf{u}^{\mathrm{ca}}$ and kinematic viscosity ν^{ca} onto their physical counterparts. From the macroscopic flow equations for the FCHC-lattice gas in the incompressible limit, i.e.

$$\begin{aligned} \rho^{\mathrm{ca}} \partial_{t^{\mathrm{ca}}} \mathbf{u}^{\mathrm{ca}} + g \rho^{\mathrm{ca}} \mathbf{u}^{\mathrm{ca}} \cdot \nabla^{\mathrm{ca}} \mathbf{u}^{\mathrm{ca}} &= -\nabla^{\mathrm{ca}} p^{\mathrm{ca}} + \rho^{\mathrm{ca}} \nu^{\mathrm{ca}} \Delta^{\mathrm{ca}} \mathbf{u}^{\mathrm{ca}} \\ \nabla^{\mathrm{ca}} \cdot \mathbf{u}^{\mathrm{ca}} &= 0 \end{aligned} \tag{8}$$

we can now derive the macroscopic flow equations in terms of the real physical variables (with special care for the calculus concerning the differential operators ∇ and ∂_t):

$$\begin{aligned} \frac{\tau}{\alpha_\rho \alpha_u} \rho \partial_t \mathbf{u} + \frac{g l}{\alpha_\rho \alpha_u^2} \rho \mathbf{u} \cdot \nabla \mathbf{u} &= -\frac{l}{\alpha_p} \nabla p + \frac{l^2}{\alpha_\rho \alpha_\nu \alpha_u} \rho \nu \Delta \mathbf{u} \\ \frac{l}{\alpha_u} \nabla \cdot \mathbf{u} &= 0 \end{aligned} \tag{9}$$

Three of the unknown coefficients $\tau, l, \alpha_\rho, p_0, \alpha_p, \alpha_u$ and α_ν follow from the requirement that the latter equation should analytically boil down to (a scalar

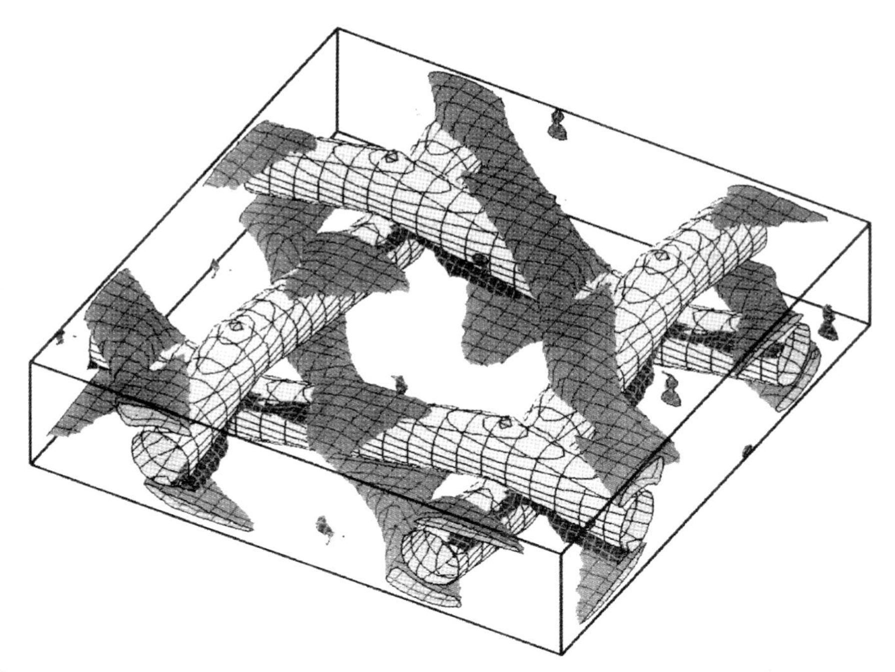

Fig. 5. Complex three-dimensional flow. The white surfaces bound a periodic geometry of woven strings, satisfying stick boundary conditions. The flow is driven by a horizontal body force directed along the diagonal from the far corner of the box towards the front. Stick boundary conditions also apply at the top and bottom faces of the box. The gray surface connects points with equal value of the diagonal velocity component, indicating the trajectory of the main flow. The small black surfaces near the crossings of the strings indicate counter-current flow. The Reynolds number of the flow based on the height of the box and the superficial velocity is 310.

multiple of) the Navier-Stokes equations.

$$\begin{aligned} l &= \frac{g\alpha_\nu}{\alpha_u} \\ \tau &= \frac{g^2\alpha_\nu}{\alpha_u^2} \\ \alpha_p &= \frac{\alpha_u^2\alpha_\rho}{g} \end{aligned} \tag{10}$$

Note that these relations capture the elimination of the g-factor from the convective term. This scaling also reveals the Reynolds number of a lattice gas simulation, obviously independent of the interpretation in real physical variables.

$$Re = \frac{uL}{\nu} \tag{11}$$

$$= \frac{g u^{\mathrm{ca}} L^{\mathrm{ca}}}{\nu^{\mathrm{ca}}}$$

Now remember that the viscosity ν^{ca} of the lattice gas is fixed by the state transition rules, and also the density ρ^{ca} is chosen at a value which is optimal with respect to computational efficiency. For most applications the density and the viscosity of the target fluid are known a priori, which determines the coefficients α_ρ and α_ν. The level of the pressure p_0 is irrelevant for the flow which leaves us only one parameter to play with in the scaling, viz. α_u.

For the sake of computational efficiency one would like to choose α_u as high as possible such that the length scale of a lattice unit is maximized. However, only low Mach number velocities are allowed in a lattice gas automaton simulation in order to recover the Navier-Stokes equations in the incompressible limit. It is common practice to keep the Mach number of a lattice gas automaton flow well below 0.4.

From the above analysis one could conclude that the mapping of the parameters of a flow experiment onto the lattice gas variables is completely determined by the properties of the state transition rules and the objective of computational efficiency. This is true if models with moderate viscosities are considered only.

Another constraint of the scaling exercise is that the contribution of $\mathcal{O}((\nabla \mathbf{u})^2)$ terms in the macroscopic flow equations should be negligible. McNamara et. al. have explained how for a given flow experiment with a given Reynolds number a critical lattice gas viscosity can be derived below which the maximum length scale of a lattice unit is limited by the requirement of velocity resolution rather than the Mach number of the flow [18]. Hence, if the viscosity of the lattice gas model is lower than this critical viscosity one should further restrict the maximum velocity in the lattice gas simulation. Experimental evidence indicates that FCHC-models with $R_* \leq 10$ can still resolve peak velocities of Mach 0.4, even if sharp features are present [19].

Satisfying the requirement of velocity resolution by lowering the velocity, is not a very popular exercise. First, the reduction of the velocity affects the computational efficiency of the simulation. Suppose that we double the velocity resolution by means of decreasing the velocity and increasing the number of lattice units in the characteristic length both by a factor of $\sqrt{2}$ (leaving the Reynolds number invariant). From equation (10) we learn that as a result the τ-coefficient will be decreased by a factor of 2! Hence, the computational effort that is needed to follow a certain feature of the flow will be increased by a factor of $(\sqrt{2})^3 \times 2 = 4\sqrt{2}$. Secondly, note that a reduction of the velocity in a lattice gas simulation severely deteriorates the signal/noise ratio in the solution. Actually, the Mach number is also bounded from below below in order to resolve the flow with at least some accuracy! If noise is an issue, we have to maintain the superficial velocity. Doubling the velocity resolution will then require twice the number of lattice units in the characteristic length and a different set of state transition rules with a twofold viscosity. In the latter case the computational cost will grow even by a factor of 16.

In fact, the limitations from noise, incompressibility, and velocity resolution impose in fact an upper bound to the velocity which is ultimately proportional to the viscosity of the model. This is illustrated by the empirical bound, which quantitatively expresses the computational cost (i.e. the number of lattice units in the typical length scale) of a lattice gas automaton simulation as a function of the Reynolds number based on the maximum speed.

$$L^{\mathrm{ca}} \geq \frac{Re}{4} \tag{12}$$

This bound is quite fundamental in the current framework of lattice gases. We have argued above, that this constraint cannot be relaxed by the possible invention of lattice gas models with still lower viscosities. A significant improvement may be achieved however, if a satisfactory subgrid model can be incorporated in the scheme. A similar discussion about accuracy, resolution and noise has been presented in the section "Too much vs. too little Resolution" of [12].

It is noteworthy, that the constraint of velocity resolution is also an issue in floating point finite difference schemes. The difference between the two techniques is, that the finite difference scheme will 'blow up' if the viscosity is set too low, while the lattice gas automaton will deteriorate the equations which are being solved.

A significant distinction between the boolean lattice gas scheme and a floating point finite difference scheme, is in their computational and storage requirements. The lattice gas technique needs only 24 bits to store all variables at a single node and only a few integer operations are used to update these variables. In return, a lattice gas simulation will require a relatively large amount of grid points to trace the flow. The net benefit from this is that a fine grid is available to discretize irregular boundary geometries. Techniques are available to incorporate various types of boundary conditions consistently with the lattice gas scheme [20]. The general approach is to apply a small computational effort in order to perturb individual states of nodes at the boundary, thereby maintaining a condition on the macroscopic scale.

The potential of a fine grid is illustrated in our example flow. The box in Figure 5 corresponds to a FCHC lattice of $1056 \times 1056 \times 240$ grid points. Actually, we only simulate one quarter of that box as we are capable of imposing mirror-periodic boundary conditions in the x- and y-directions. All particles which move through one of the four sides of the box will have their z-coordinate and z-velocity component reflected in the xy-plane. The solid strings are discretized at the node level in this lattice and the well-known 'bounce back' transition rules will impose a stick-boundary condition at their surface. Macroscopic averages of the velocity profile and the density field have been taken over volumes of size $16 \times 16 \times 8$, yielding a resolution of $33 \times 33 \times 30$ data points in one quarter of the box. Figure 5 shows contour surfaces of the component of the flow velocity that is parallel to the line $x = y$. The gray contour connects points with a diagonal velocity component of $0.77 m/s$, the black surface near the crossing of the strings reveals counter current flow with a diagonal component of $-0.012 m/s$.

The flow was driven by a uniform horizontal body force **f** directed along the $x = y$ diagonal. Effectively such a body force replaces a linear pressure drop along the same diagonal direction. In our simulation we had to apply $f = 0.000025$ flips per node per time step both in the $x-$ and in the $y-$direction in order to maintain the constant superficial flow of 0.0806 lattice units per time step (i.e. $0.31m/s$ after scaling). This flip rate predicts an average pressure drop of $1.17bar/m$ along the diagonal direction according to

$$-\nabla p = \frac{2\alpha_p}{l} f \qquad (13)$$

Obviously, the $33 \times 33 \times 30$ matrices representing time-dependent solution of the pressure and velocity profile provide a richness of data that can be analysed from many perspectives. We will not discuss all features of this particular example flow, but rather present just another characteristic. Figure 6 shows a contour surface of the vorticity profile that is generated by the geometry. The stretching of vorticity downstream can clearly be observed. The vorticity is calculated by numerically differentiating the velocity profile. More detail of the structure of the flow near a crossing of the strings is depicted in Figure 7. The two vortices that are visible in this picture are trapped between the geometry and the relatively strong main stream as depicted in Figure 5. The time-dependent flow profile of our example does not show a clear shedding of large vortex structures. However, completely different flow profiles do emerge when the direction of the pressure drop is changed, or the geometry is altered!

5 Conclusions

In the past years, the lattice gas automaton technique has evolved into a numerical tool that is capable of solving complicated flow problems. Quite some characteristics of the technique are shared with a conventional finite difference scheme, i.e. the three-dimensional Navier-Stokes equations are fully resolved and Reynolds numbers are limited to moderate values ($Re < 1000$). However, there are some important differences. The lattice gas technique does not suffer from problems of numerical stability, as only integer arithmetic is used. Instead, its solutions are inherently noisy, which limits the resolution of the solution. The lattice gas technique can deal with complicated boundary conditions and geometries, as space is discretized on a fine grid.

In this paper a new algorithm has been introduced for actually running the lattice gas automaton model. With this algorithm we now can exploit efficiently the three-dimensional FCHC models with high 'quality' (R_*), which have been introduced by various authors in the past few years. The algorithm has been implemented on our parallel computer with 400 transputers, but it can be run as easily on an ordinary workstation.

In the last section of this report we have illustrated the potential of this new simulation tool, by presenting the early results of the computation of a three-dimensional flow through a complicated net of strings. Currently, there is

Time Dependent Anomalous Diffusion for Flow in Multi-fractal Porous Media *

Frederico Furtado **, *James Glimm* ***, *Brent Lindquist* **, *Felipe Pereira* **, *and Qiang Zhang* **

Department of Applied Mathematics and Statistics,
University at Stony Brook, Stony Brook, NY 11794-3600

1 Introduction

Macroscopic flow computations are used in the modeling of most practical porous media flow studies. These computations require macroscopic or effective flow parameters, describing effective fluid behavior over an integrated range of shorter length scales. This requirement is the primary reason for solving the scale up problem, which is that of determining or extrapolating such effective parameters from data measured at smaller length scales. Multi-length scale characterization of the medium heterogeneity is needed to solve this scale up problem. Such a characterization is usually missing or incomplete. Thus a second value of solving the scale up problem is to specify a minimum level of knowledge of porous media heterogeneity required in order to narrow the range of uncertainty in flow computations associated with an incomplete description of the media. We believe a solution of the scale up problem, adequate for these two purposes, is possible, and is one of the goals in our current investigations.

We concentrate on the effective diffusivity as a macroscopic model parameter describing the growth rate of a fluid mixing zone induced by a heterogeneous permeability field. The effective diffusion process can be non-Fickian (anomalous) for two reasons: multi-fractal rock statistics producing non-Fickian transient effects, or fractal rock statistics with slowly decaying correlations producing a non-Fickian, steady state response. The present article analyses the transient effects as a cause of non-Fickian diffusion, while earlier studies, [AHLL], [FGLP1], [FGLP2] and [GS] concerned the steady state response to fractal statistics. For the transient response work, we follow the theoretical analysis of [Z].

* Support from the computational facilities of the Numerical Aerodynamics Simulations Systems Division, NASA Ames Research Center, Moffett Field and of the Engineering Physics and Math Division, Oak Ridge National Laboratory is gratefully acknowledged.

** Supported by the Applied Mathematics Subprogram of the U.S. Department of Energy DE-FG02-90ER25084.

*** Supported by the Applied Mathematics Subprogram of the U.S. Department of Energy DE-FG02-90ER25084 and the National Science Foundation, grant DMS-89018844.

We define multi-fractal behavior as the modification of self-similar (fractal) behavior to account for different events occurring with different probabilities on different length scales. In section 2 we demonstrate the inclusion of this behavior in our earlier fractal models by considering $\xi = \log(\text{permeability})$, and its covariance function

$$C(\mathbf{x},\mathbf{y}) \equiv <\xi(\mathbf{x})\xi(\mathbf{y})>= C(r) \ , \qquad r \equiv |\mathbf{x}-\mathbf{y}|, \tag{1}$$

which we assume depends only on separation distance r (i.e., we assume stationarity and isotropy of the field ξ).

We note that further generalizations of our hypotheses are important, and will be addressed in later papers in this series. In order to understand the importance of transverse flow in the mixing process we will study the fluid flow in partially layered media. The inclusion of bimodal geology, such as shale barriers, would require the addition of Ising- or ϕ^4-model statistics [GJ] to the Gaussian models we have used. The influence of curvilinear flow geometry is very important and previous work of the authors [GILMY] should be extended to the present context. Finally nonlinear flow physics (immiscible flow) should also be considered. The influence of microscopic mixing (molecular or capillary diffusion) and the interaction between macroscopic and microscopic diffusion (Taylor diffusion) appear to have smaller effects in comparison with the above list.

2 Multifractal Diffusion

We begin from a microscopic, linear transport equation coupled to a random velocity field,

$$s_t + \mathbf{v}\cdot\nabla s = 0 \ . \tag{2}$$

The variable s denotes the volume fraction (saturation) of a tagged fluid displacing its untagged (saturation $1-s$) counterpart. We take the velocity field, $\mathbf{v}$, to be a random field, obtained by solution of Darcy's law

$$\mathbf{v} = -\frac{\overline{K}\, e^{\xi}}{\mu}\,\nabla P \ , \tag{3}$$

and the incompressibility condition,

$$\nabla\cdot\mathbf{v} = 0 \ . \tag{4}$$

In (3), μ is the viscosity of the fluid, P is the pressure, $\overline{K}\, e^{\xi}$ is the rock permeability, which we assume to be a scalar field, $\overline{K}$ is constant, and ξ is a random field, with stationary Gaussian statistics and mean value $<\xi>= 0$, describing the statistical variation of the rock permeability.

We choose boundary conditions for (2) -- (4) so that the dominant flow is in the y-direction. Expanding (3) in powers of ξ, we obtain, to first order

$$\mathbf{v} = \mathbf{v}_0 + \delta\mathbf{v} \ , \qquad \delta\mathbf{v} = \xi\mathbf{v}_0 - v_0\nabla\left(G * \frac{\partial\xi}{\partial y}\right) \ , \tag{5}$$

where $\mathbf{v}_0 = v_0 \mathbf{e}_y$ is the unperturbed, constant, y-direction flow, and $G(\mathbf{x}, \mathbf{x}')$ is the Green's function for the laplacian

$$\Delta G(\mathbf{x}, \mathbf{x}') = \delta(\mathbf{x} - \mathbf{x}') \ .$$

Thus, to first order in ξ, the velocity field is also Gaussian. However, with boundary conditions imposed on a finite length domain in the y-direction, $\delta\mathbf{v}$ will not satisfy stationary statistics, even if ξ does. We, therefore, simplify our formulas by assuming that the boundaries are at infinity, in which case $\delta\mathbf{v}$ also satisfies stationary statistics. This first order theory will, of course, be exact in the limit of weak variation in the heterogeneity ξ. One of the conclusions of our numerical studies is that the formulas we have derived for mixing length exponents in this weak heterogeneity limit are valid up to heterogeneities of moderate strength.

Flow under (2) – (4) results in macroscopic mixing induced by the heterogeneity in the rock permeability (and hence in $\mathbf{v}$). Assuming dominant flow in the y-direction, using (5), performing an ensemble average over all rock realizations having the same statistics and a spatial average over directions transverse to the y-direction, the resultant averaged saturation $< s >$ satisfies [D], [Z] a 1-D macroscopic advection-diffusion equation

$$\frac{\partial < s >}{\partial t} + v_0 \frac{\partial < s >}{\partial y} = D(t) \frac{\partial^2 < s >}{\partial y^2} \ , \tag{6}$$

where $D(t)$ is an effective, time dependent viscosity coefficient. The derivation of (6) from (2) - (5) is exact to second order in the perturbation expansion in ξ.

Under the above considerations, the time dependent, effective diffusion parameter $D(t)$ can be shown [D], [Z] to be given, to leading order, by

$$D(t) = \oint_0^t < \delta\mathbf{v}(\mathbf{x} - \sigma \mathbf{v}_0) \ \delta\mathbf{v}(\mathbf{x}) > d\sigma \ , \tag{7}$$

where the integral is along backward streamlines of the unperturbed velocity field $\mathbf{v}_0$.

Consider first the time-asymptotic behavior $D_\infty \equiv \lim_{t\to\infty} D(t)$. If D_∞ is finite, the asymptotic large time diffusion process is Fickian (normal), whereas if $D_\infty = \infty$, the asymptotic large time diffusion process is non-Fickian (anomalous) [Z]. An examination of (5) to lowest order in ξ, shows that the diffusion will be anomalous if and only if the backward streamline integral, analogous to (7), for the rock correlation $< \xi \ \xi >$ diverges.

To analyze finite time (transient) behavior, assuming stationary isotropic statistics, we express the correlation functions $< \xi \ \xi >$ and $< \delta\mathbf{v} \ \delta\mathbf{v} >$ (along streamlines) in multi-fractal form [Z]

$$< \xi(0) \ \xi(r) >= b(r) \, r^{-\beta(r)} \ , \qquad < \delta\mathbf{v}(0) \ \delta\mathbf{v}(r) >= a(r) \, r^{-\rho(r)} \ ,$$

where $\beta(r)$ (similarly ρ) is the slope and $b(r)$ (similarly a) is the intercept of the tangent line $\tau(r)$ of the graph $\ln(< \xi \ \xi >)$ (similarly $\ln(< \delta\mathbf{v} \ \delta\mathbf{v} >)$) versus $\ln(r)$. This multi-fractal representation is useful when $\beta(r)$ and $\rho(r)$ are *slowly* varying functions, which we now assume. The effective saturation variable $< s >$ develops a mixing length $l(t)$ which can also be represented in multi-fractal form,

$$l(t) \sim t^{\gamma(t)} , \qquad \gamma(t) = \frac{1}{2} \frac{\partial \ln D(t)}{\partial \ln(t)} + \frac{1}{2} , \tag{8}$$

where, for slowly varying β (and ρ),

$$\gamma(t) = \max\{\frac{1}{2}, 1 - \frac{\beta(t)}{2}\} . \tag{9}$$

Finally, for initial growth behavior, assuming the $< \xi \; \xi >$ correlation function is nonsingular at short distances, the initial mixing length exponent has been determined [Z] to be $\gamma(t = 0) = 1$. Thus we see that, assuming the flow is asymptotically Fickian, $\gamma_\infty = \lim_{t\to\infty} \gamma(t) = 1/2$, the transients in the mixing process are necessarily non-Fickian and that the multi-fractal theory applies to the transient flow.

3 Computational Results

We report on two dimensional computational studies, using the front tracking method [GIMM] [GLMP] for high resolution of the mixing zone, for three cases of rock permeability exponent: $\beta = 0.5$, $\beta = 1.25$ and $\beta = \infty$.

For each case, specific statistical realizations of the rock heterogeneity fields were generated by convolution of an independent Gaussian random field (white noise) with a weight function [FGLP1]. For $0 < \beta < 2$, the weight function has a power law decay, while in the case $\beta = \infty$ the weight function is a δ function, which we smear out over one mesh block. We note that this numerical rock permeability field, ξ, is not purely fractal, but is cut off at short distance by the effect of finite grid spacing. For each separate value of β, the results discussed below are obtained from averaging over a number of realizations generated on grids of the same size. The number of realizations and grid sizes used for each value of β in the studies discussed below are given in Table 1.

For a given realization of the permeability field, the magnitude of the heterogeneity can be quantified by the coefficient of variation

$$C_v \equiv \frac{\sigma'}{< K' >},$$

where σ' and $< K' >$ are, respectively, the standard deviation and mean of the permeability values in the given realization. The average coefficient of variation obtained from all realizations generated for a given β and grid size is given in column four of Table 1. The uncertainty quoted for each average indicates the range of coefficients of variation found in each of the individual realizations which contributed to the average.

In Fig. 1 we plot the numerically computed $\ln(< \overline{K}e^\xi, \overline{K}e^\xi >)$ correlation as a function of $\ln(r)$ for the slowest decaying correlation fields, $\beta = 0.5$ having average $C_v = 0.19$. The correlation displays pure fractal behavior over a distance

Table 1. Number of realizations, grid sizes and coefficient of variations used for each value of β

β	number realizations	grid size	C_v average
0.5	5	128 x 512	0.186 ± 0.009
1.25	25	32 x 512	0.189 ± 0.002
∞	12	32 x 256	0.182 ± 0.004
∞	12	32 x 256	0.38 ± 0.01
∞	12	32 x 256	0.69 ± 0.02
∞	12	32 x 256	0.96 ± 0.04

of about 55 mesh blocks. The slope of this curve is 0.47, in 6% agreement with the theoretical decay exponent. Beyond this distance (not shown), the correlation falls off more rapidly and is not fractal. We believe that lack of statistical convergence is the cause of such large distance behavior, since fewer statistics are accumulated in a finite size domain when computing the correlation for large r separation than for short separations.

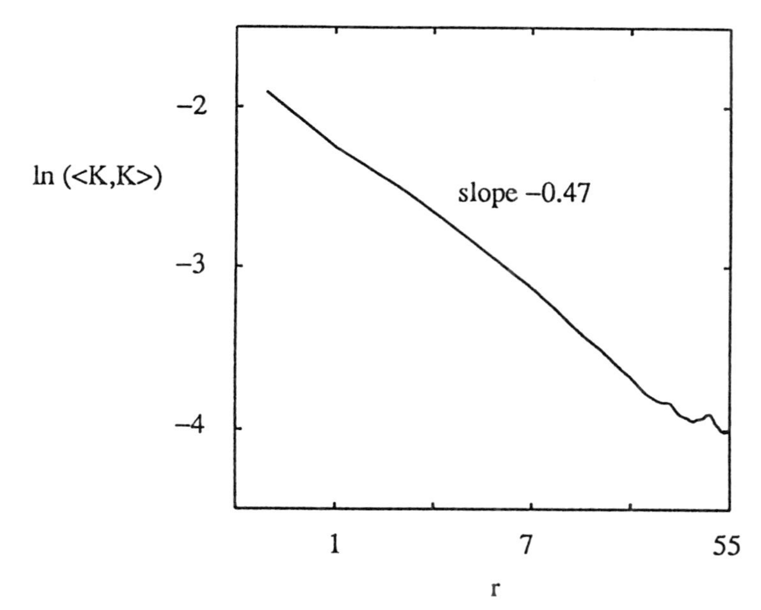

Fig. 1. Correlation function $\ln(<\overline{K}e^{\xi(r)}\overline{K}e^{\xi(0)}>)$ versus $\ln(r)$ for $\beta = 0.5$, $C_v = 0.19$.

Because of their rapid decay, the permeability correlation functions for $\beta = 1.25$ and $\beta = \infty$ are harder to measure on grids of finite spacing. For $\beta = \infty$, the measured correlation function is consistent with a numerical δ function, dropping to 1% of its zero separation value in one mesh block distance. The correlation function for $\beta = 1.25$ drops to 1% of its zero separation value over four mesh blocks. The *zero* separation value is based on a subgrid resolution distance corresponding to 1/32 of a mesh block width. This was determined from local mesh refinement used in generating the permeability field.

For fixed values of β, C_v and grid size, the mixing length $l(t)$ was determined from the ensemble and spatially transverse averaged profiles of $< s >$. At a fixed t, the profile was fit with a complementary error function erfc containing two free parameters, a mean profile travel distance L (it is convenient to use the travel distance L rather than the time t), and a variance, whose square root determines the mixing length $l(L)$. In Fig. 2 we display the averaged saturation profiles $< s >$ and fitted erfc profiles, at similar L values, as a function of C_v obtained for $\beta = \infty$ fields. For the largest C_v (i.e. the greatest heterogeneity strength) the profiles begin to deviate from an erfc. It is not clear whether averaging over more realizations would be sufficient to correct this deviation at large heterogeneity strengths.

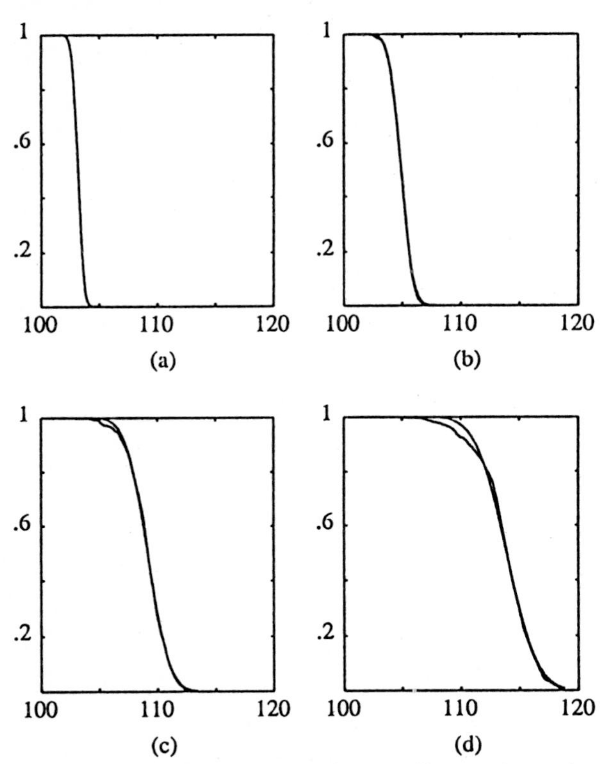

Fig. 2. The averaged saturation profiles at approximately fixed average flow distance L for flows in $\beta = \infty$ permeability fields characterized by 4 different coefficients of variation: (*a*) 0.18 ± 0.005, (*b*) 0.38 ± 0.01, (*c*) 0.69 ± 0.02 and (*d*) 0.96 ± 0.04. Superimposed upon each is the best fit erfc (smooth curve).

Expressing the asymptotic mixing length behavior $l_\infty \equiv l(L \to \infty)$ in pure fractal form

$$l_\infty = a_\infty t^{\gamma_\infty} ,$$

it can be shown that, in the small ξ limit, the fractal exponent γ_∞ is independent of C_v, and the scale factor a_∞ varies linearly with C_v. The results of mixing length exponent γ obtained for $\beta = \infty$ from four separate sets of runs (corresponding to rows 3 to 6 of Table 1) with differing C_v are shown in Table 2. The results are all within 2% agreement of the asymptotic prediction of $\gamma_\infty = 0.5$ except for the largest C_v run which has 12% error. These results were obtained by fitting the mixing length growth over the range $7 \leq L$. Thus it appears that, for $\beta = \infty$ and $C_v < 0.96$, the asymptotic growth rate is achieved within a distance of 7 mesh blocks.

Table 2. Mixing length exponent γ for $\beta = \infty$ for four sets of runs with different coefficients of variation C_v.

C_v	0.18 ± 0.005	0.38 ± 0.01	0.69 ± 0.02	0.96 ± 0.04
γ	0.50	0.49	0.49	0.56

In Fig. 3 we plot a_∞ versus C_v for $\beta = \infty$ as measured from the runs in Table 2. The variation does indeed appear linear, except for the largest C_v used, suggesting that weak field theory is no longer appropriate above this value.

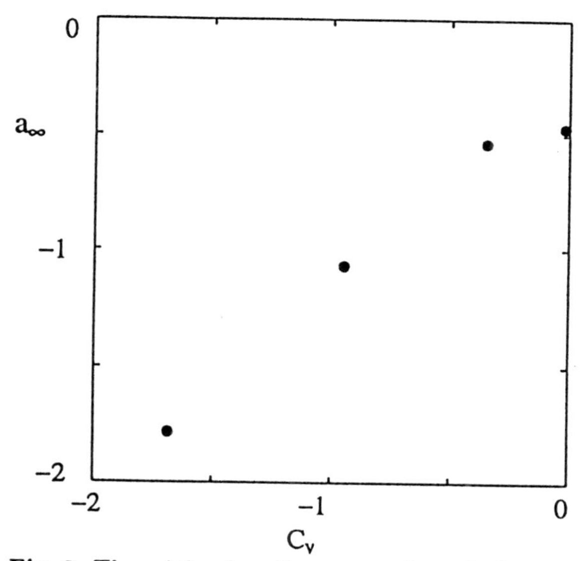

Fig. 3. The mixing length asymptotic scale factor a_∞ versus C_v for the field $\beta = \infty$.

To check numerical convergence of the fluid flow/mixing length computations, a mesh refinement study was performed for the $\beta = \infty$, $C_v = 0.38$ permeability field data (line 4 of Table 1). Fig. 4a shows the mixing length growth obtained on a fluid grid (32 x 256) equal in size to the grid used to generate the permeability data. The mixing length growth obtained on a fluid grid of (64 x 512) (at the limits of the computational memory available) is displayed in Fig. 4b. The computed mixing length growth exponents, γ, agree with each other and with the asymptotic theory to within 2%.

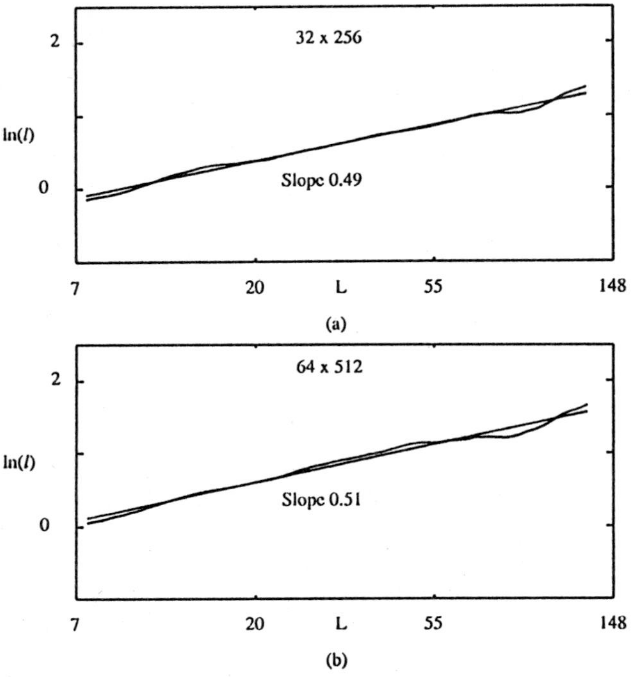

Fig. 4. $\ln(l)$ versus $\ln(L)$ for $\beta = \infty$, $C_v = 0.38$ computations obtained on flow grids of a) 32 x 256 b) 64 x 512 .

The $\beta = \infty$ results above indicate behavior consistent with pure fractal, Fickian behavior (with perhaps very rapidly decaying transients), at least in the small heterogeneity strength regime $C_v < 0.9$.

Next we describe numerical simulations for two non-Fickian diffusion processes. We begin with the consideration of transient anomalous diffusion. Theoretically [Z], fluid flow in fractal permeability fields with $\beta = 1$ should exhibit the slowest decay to asymptotic growth rate. Thus, of our three sets of runs, $\beta = 1.25$ should exhibit the most noticeable transient effects.

Figure 5a plots mixing length growth versus distance for the $\beta = 1.25$ case listed in Table 1. Superimposed is a least-squares straight line fit, of slope 0.57, revealing the curvature of the data. In Fig. 5b, the $\ln(r)$ axis is broken into thirds, each third fitted by a least-squares straight line. The slopes of these successive data fits reveal the transient decay towards the predicted asymptotic value of

$\gamma_\infty = 0.5$. In [Z], a correlation of the form

$$< \delta v(0)\delta v(r) >= \frac{a}{(b+r)^{\beta_\infty}} \tag{10}$$

is chosen to provide a simple interpolation from short distance $\gamma(0) = 1$ behavior, to long distance γ_∞ asymptotics for the mixing length. Velocity correlations of the form (10) result, to leading order, from rock correlations of the same form, as can be seen from (5). (The second term on the right hand side has a much more rapid decay than the first for $\beta < 2$.) Using (10), with a, b as free parameters and $\beta_\infty = \beta(\infty) = 1.25$, we fit the computed diffusion length data. The fitted profile, compared with the computational data, is displayed in Fig. 6. The best fit parameters obtained are $a = 3.8 \cdot 10^{-5}$, $b = 0.27$.

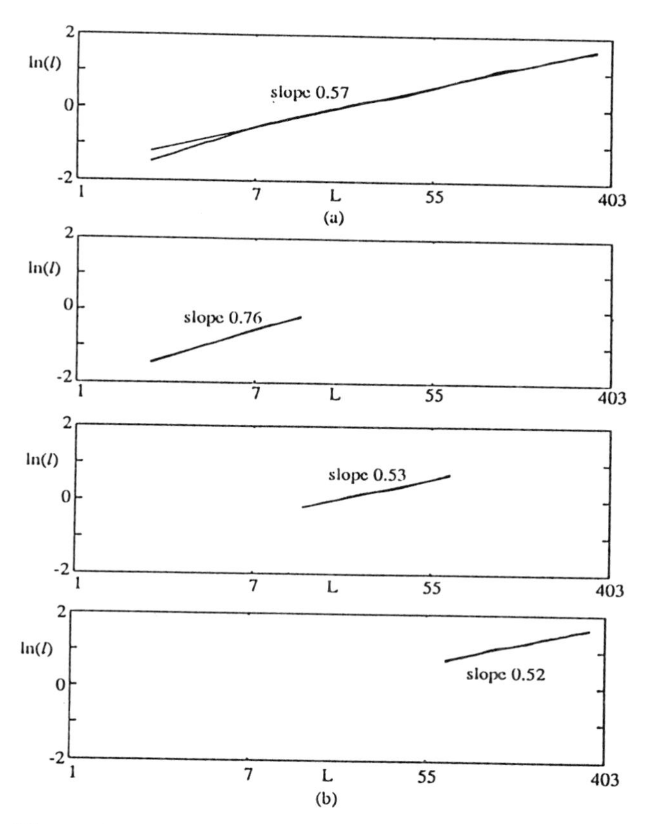

Fig. 5. Behavior of $\ln(l)$ versus $\ln(L)$ for the $\beta = 1.25$ runs. (a) Straight line fit shows non-fractal effects at short distance. (b) Straight line fits to successive thirds of the data exhibit evidence of a transient approach to a limit of 0.5.

Finally, we discuss anomalous diffusion generated by slowly decaying fractal rock statistics. Figure 7 displays the mixing length growth for the $\beta = 0.5$ case listed in Table 1. Omitting the range $L < 7$, we find the data well fit by a straight

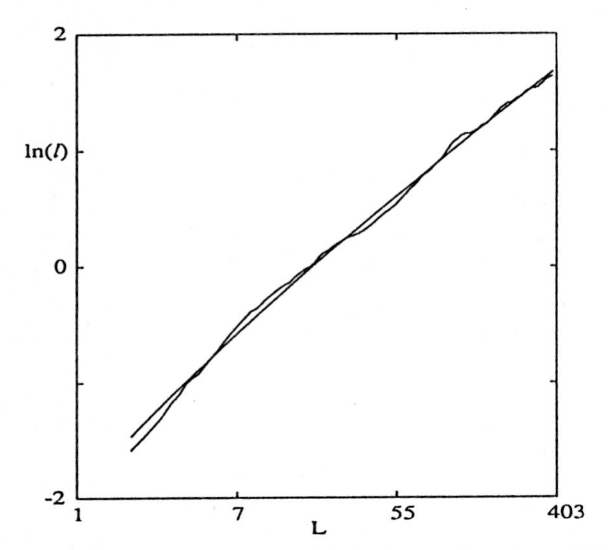

Fig. 6. Two parameter, multi-fractal fit based on (10) to the mixing length growth data for the $\beta = 1.25$ runs. Best fit parameters are $a = 3.8 \cdot 10^{-5}$, $b = 0.27$.

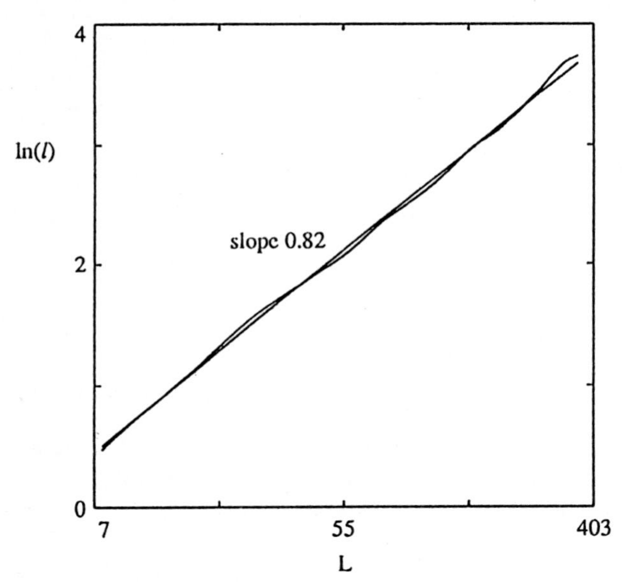

Fig. 7. $\ln(l)$ versus $\ln(L)$ for the $\beta = 0.5$ case.

line having slope $\gamma = 0.82$, in 9% agreement with the theoretical asymptotic limit of 0.75. This behavior is consistent with non-Fickian fractal response. Studies such as performed in Figs. 5 and 6 for the $\beta = 0.5$ data do not indicate long time transient decay similar to that seen in the $\beta = 1.25$ data.

4 Discussion and Conclusions

We have described a new theory which provides quantitative predictions for the growth rate of the mixing length in the case of multi-fractal heterogeneity functions. This theory is supported by our present computational studies, except for very large permeability heterogeneity strength. We identify three distinct growth types dependent on the fractal exponent for the correlation function of the log-normal rock permeability: Fickian (normal) steady state growth, non-Fickian transient behavior slowly decaying to asymptotic (either Fickian or non-Fickian) rate, and non-Fickian steady state growth.

References

[GJ] Glimm, J., Jaffe, A.: Quantum Physics: A Functional Integral Point of View. Springer-Verlag New York (1987)

[GS] Glimm, J., Sharp, D.: A Random Field Model for Anomalous Diffusion in Heterogeneous Porous Media. J. Stat. Phys. To appear

[FGLP1] Furtado, F., Glimm, J., Lindquist, B., Pereira, F.: Multi-Length Scale Computations of the Anomalous Mixing Length Growth in Tracer Flow. Proceedings of the Emerging Technologies Conference F. Kovarik Houston, TX, (to appear)

[FGLP2] Furtado, F., Glimm, J., Lindquist, B., Pereira, F.: Characterization of mixing length growth for flow in heterogeneous porous media. SPE paper # 21233, Proceedings of the 11th SPE Symposium on Reservoir Simulation (Feb. 1991)

[Z] Zhang Q.: A Multi-Length Scale Theory of the Mixing Length Growth for Tracer Flow in Heterogeneous Porous Media. In preparation

[GILMY] Glimm, J., Isaacson, E., Lindquist, B., McBryan, O., Yaniv, S.: Statistical Fluid Dynamics II: The Influence of Geometry on Surface Instabilities. Frontiers in Applied Mathematics **1** SIAM Philadelphia (1983)

[D] Dagan G.: Solute transport in heterogeneous porous formations. J. Fluid Mach. **145** (1984) 151–177

[AHLL] Arya, A., Hewett, T., Larson, R., Lake, L.: Dispersion and reservoir heterogeneity. SPE Res. Eng. **3** (Feb, 1988) 139–148

[GIMM] Glimm, J., Isaacson, E., Marchesin, D., McBryan, O.: Front tracking for hyperbolic systems. Adv. Appl. Math., **2** (1981) 91–119

[GLMP] Glimm, J., Lindquist, B., McBryan, O., Padmanhaban, L: A front tracking reservoir simulator: five-spot validation studies and the water coning problem. Frontiers in Applied Mathematics bf1 SIAM, Philadelphia (1983)

This article was processed using the LaTeX macro package with ICM style

Porous Media Flow on Locally Refined Grids

R. H. J. Gmelig Meyling, W. A. Mulder, and G. H. Schmidt

Royal/Dutch Shell, Exploration and Production Laboratory,
P.O. Box 60, NL-2280AB Rijswijk, The Netherlands

Abstract

A numerical method is described for the simulation of miscible and immiscible fluid flow in hydrocarbon reservoirs. A flexible gridding technique introduces both static and dynamic local grid refinement for multiple space dimensions. Operator splitting and defect correction are used to deal separately with elliptic and hyperbolic problems. The elliptic equations are discretised by mixed finite elements and solved by multigrid. Mixed finite elements provide an accurate representation of the flows and allow for a good connection with the hyperbolic equations. The method of characteristics is used for the hyperbolic problems, eliminating numerical diffusion almost completely. Nonlinear phenomena are treated by suitable Riemann solvers.

1 Introduction

The simulation of fluid flow through porous media is of crucial importance in hydrocarbon reservoir engineering. This paper presents some aspects of the research on numerical methods for porous media flow at the Royal Dutch/Shell Laboratory.

A number of features of the computational method will be described. These features must be compatible so that they can be combined into a solution method both for miscible and immiscible fluid displacement. The starting point of the computations is the set of equations for fluids consisting of N (chemical) components in M phases, with N and M arbitrary. A review of the modelling aspects of these equations can be found e.g. in Allen et al. (1988).

The first feature concerns the gridding. The highly varying length scales occurring in large-scale reservoir simulation give rise to a lasting demand for *flexible gridding*. The flow near wells and geological heterogeneities needs locally high accuracy. A fine grid moving along with the solution is required to avoid numerical diffusion at sharp (saturation) fronts and to compute their position accurately. The literature on dynamic gridding distinguishes methods with moving gridpoints (see e.g. Miller (1986)) and methods based on dynamic refinement in a grid with static points (see Ewing (1988) for an excellent review). We use the latter method, since we wish to perform computations in up to three space dimensions. Moreover, we want a uniform approach for one, two, and three space dimensions. More details of this gridding technique can be found in Schmidt (1990).

The second feature concerns the discretisation. Many of the equations are discretised by using *mixed finite elements.* This means that the fluid fluxes are primary variables, instead of eliminating them and using only scalar fields (such as pressures) as primary variables. Many computational techniques for porous medium flow are based on this discretisation (see e.g. Chavent et al. (1984), Russell and Wheeler (1983)). These authors all refer back to Raviart and Thomas (1977). We use this discretisation, not only because of accuracy arguments, but also because it is suitable for introducing local refinement and for imbedding the use of characteristics into the method. For the combination of mixed finite elements and characteristics see also Douglas and Yirang (1988).

The third feature concerns the solution of large systems of coupled equations. *Multigrid* is a technique to solve very large systems and is used in many areas of computational fluid dynamics, see e.g. Wesseling (1990). The role of multigrid in our approach is to solve symmetric systems only. The technique is specially adapted to the mixed finite elements; for details see Schmidt and Jacobs (1988).

The last feature is the use of the *method of characteristics* to solve hyperbolic systems of equations. As compared to standard finite differences and finite elements, this introduces almost no numerical diffusion and permits very large time steps. It is common in reservoir simulation to use characteristics to solve the two-phase Buckley-Leverett equations in one-dimensional problems. In this paper, characteristics are used to solve hyperbolic equations in up to three space dimensions. Riemann solvers enable us to treat nonlinear effects and to maintain a large time step (see Gmelig Meyling (1990)).

The plan of this paper is as follows. Section 2 describes the general equations for porous media flow. The solution techniques are presented in Sect. 3, whereas Sect. 4 outlines the adaptive gridding. Although the numerical techniques are quite general, they have so far only be applied to *two*-phase flow. Numerical examples for the immiscible flow of oil and water are presented in Sect. 5.

2 Equations for porous media flow

For the flow of fluid consisting of N (chemical) components in M phases through a porous medium we have the following relations.

First, there is a conservation law for each component (chemical reactions are not considered in this paper):

$$\frac{\partial a_n}{\partial t} + \nabla \cdot \mathbf{u}_n = q_n\,, \quad n = 1, \ldots, N \tag{1}$$

Second, the flux of each phase is given by Darcy's law:

$$\mathbf{v}_m = -\lambda_m \mathcal{K} \left(\nabla p_m - \rho_m \mathbf{g} \right), \quad m = 1, \ldots, M \tag{2}$$

Third, the equations are coupled by a number of phenomena:

- The components are convected in the phases, which implies that the $\mathbf{v}_m$ occur (usually linearly) in expressions for the $\mathbf{u}_n$.
- The pressures p_m depend on the concentrations a_n.

A closed set of equations is obtained by imposing the proper initial- and boundary conditions, so that a unique solution may be expected. The features of such a full set of equations depend strongly on the specific fluids and the properties of the porous medium, but usually a combination of elliptic, parabolic and hyperbolic behaviour is observed.

3 Solution techniques

Observation of the flow through porous media leads to distinction into (i) fields which have a strong spatial coupling (e.g., pressure and total fluid flow), and (ii) fields which have (in finite time-intervals) limited ranges of mutual dependence (e.g., saturations and concentrations). *A good solution technique should incorporate this distinction.* For a broad class of problems in porous media flow, we can reduce the full set of PDE's to a basic elliptic set of PDE's and a basic hyperbolic set of PDE's, by using operator splitting and defect correction. Parabolic behaviour is accomodated by the elliptic equations, by virtue of implicit discretisation of the time derivative. The details of the operator splitting and defect correction depend strongly on the specific application and are beyond the scope of this paper. The two basic sets of equations are:

The *basic elliptic set of equations*:

$$cp + \nabla . \mathbf{u} = q\,, \tag{3}$$

$$\nabla p + \mathbf{W}\,\mathbf{u} = \mathbf{g}\,, \tag{4}$$

to be solved for the scalar p and the vector $\mathbf{u}$, as functions of the space co-ordinate vector $\mathbf{x}$. The scalar c usually represents a compressibility coefficient divided by the time step. The tensor $\mathbf{W}$ usually represents the inverse of a permeability or diffusion tensor.

The *basic hyperbolic set of equations*:

$$\frac{\partial a_n}{\partial t} + \nabla . \mathbf{f}_n(a_1, \ldots, a_N, \mathbf{x}) = q_n(a_1, \ldots, a_N, \mathbf{x}), \quad n = 1, \ldots, N, \tag{5}$$

to be solved for the concentrations $a_1, \ldots, a_N$ as functions of time t and space co-ordinate $\mathbf{x}$.

The solution of the full set of equations consists of a series of time steps. Each time step of size Δt requires solution of the two basic sets at least once. Efficiency and accuracy are obtained by selecting appropriate techniques for each of the basic sets.

3.1 Multigrid for elliptic equations

The technique we use for solving (3–4) has been reported on e.g. in Schmidt and Jacobs (1988). Here we mention just some of its features, mainly in relation with its role in the scheme for solution of the full set of equations.

- The equations (3–4) are discretised using *mixed finite elements.* This implies a direct approximation of $\mathbf{u}$, instead of eliminating it by equation (4). We expect that the elliptic equations are thus better imbedded in the solution scheme of the full set of equations. The coupling of (3–4) with the other (hyperbolic) equations is primarily through $\mathbf{u}$.
- We use mixed finite elements of *lowest order*, i.e., a piecewise constant representation for p and a piecewise linear representation for the components of $\mathbf{u}$ (see Raviart and Thomas (1977)). The coefficients in (3–4) change rapidly (even over one or a couple of meshwidths) in the usual applications, which makes higher order approximation less suitable. When very rapid changes in the coefficients occur (e.g., at fronts, near wells, and near geological discontinuities), accuracy is improved by means of local grid refinement.
- The discrete equations are solved by multigrid as described for 2D problems in Schmidt and Jacobs (1988). Our current codes work on 1D, 2D and 3D grids with similar performance. The role of multigrid is strictly limited to the solution of the *symmetric* system (3–4). All non-symmetric terms are dealt with in the hyperbolic equations by means of suitable operator splitting and/or defect correction.
- The multigrid process solves the discretisation of (3) always with machine-accuracy, while the accuracy of solving (4) depends on the number of multigrid iterations made. This is in accordance with the usual physical interpretation of these equations: equation (3) represents fluid conservation (i.e., *kinematics*), and equation (4) represents a law of fluid *dynamics.* The coefficients in the laws of fluid dynamics (e.g., permeability and diffusion) are known with a relatively low accuracy.

3.2 Method of characteristics for hyperbolic equations

We can discern, in principle at least, two methods to solve (5). The first method is to discretize all derivatives by finite elements or finite differences. This introduces considerable numerical diffusion and/or the time step is severely restricted by stability constraints of the *Courant-Friedrich-Lewy* type. A survey of these computational difficulties can be found in Ewing (1983). In the second method one rewrites the PDE's (5) as a set of ordinary differential equations and one solves these ODE's by appropriate techniques. This is known as the *method of characteristics.* Minimal numerical diffusion and large time steps are advantages of the second method. Hence, whenever (5) can be rewritten as a system of ODE's, one should use the second method.

Both methods can be applied on the locally refined grids to be introduced in the next section. Especially the second method profits from the high resolution given by these grids. This paper reports recent developments of the second method.

For a broad and certainly important class of flow problems in porous media the hyperbolic set of equations can be written as:

$$\frac{\partial a_n}{\partial t} + \nabla . (f_n(a_1,\ldots,a_N)\mathbf{u}(\mathbf{x})) = q_n(a_1,\ldots,a_N,\mathbf{x}), \quad n = 1,\ldots,N \tag{6}$$

with initial condition $a_n(\mathbf{x}, t = 0) = a_n^0(\mathbf{x})$ for $n = 1,\ldots,N$. Equation (6) is to be solved for a_n, $n = 1,\ldots,N$ as functions of t and $\mathbf{x}$. Recall that these equations hold during one step in the time-stepping scheme as given above. The *total fluid velocity* $\mathbf{u}$ is defined as the sum of the individual component velocities. In each time step the flow $\mathbf{u}$ is computed at least once from the elliptic system (3–4). The representation of $\mathbf{u}$ by mixed finite elements is favourable (apart from accuracy) for the integration scheme to be given below. From now on, we assume that $\mathbf{u}(\mathbf{x})$ is given.

The problem (6) is in N_d space dimensions. It can be reduced to a sequence of *one-dimensional* initial value problems of the form:

$$\frac{\partial a_n}{\partial t} + \frac{\partial}{\partial \xi} f_n(a_1,\ldots,a_N) = q_n(a_1,\ldots,a_N)\ , \tag{7}$$

$$a_n(\xi, t = 0) = a_n^0(\xi), \quad n = 1,\ldots,N \tag{8}$$

along *streamlines.* The streamlines are trajectories in $\mathbf{x}$-space defined by the system of ODE's:

$$\frac{d\mathbf{x}}{d\xi} = \mathbf{u}(\mathbf{x}). \tag{9}$$

Here the parameter ξ runs along streamlines in the direction of flow. If we replace $\mathbf{u}$ by its mixed finite element representation, the system (9) can be solved analytically block-by-block (see Gmelig Meyling (1990)).

Taking account of the piecewise constant representation of a_n, $n = 1,\ldots,N$ with respect to grid blocks turns the system (7–8) into a sequence of *Riemann problems*:

$$\frac{\partial a_n}{\partial t} + \frac{\partial}{\partial \xi} f_n(a_1,\ldots,a_N) = q_n(a_1,\ldots,a_N)\ ,\ n = 1,\ldots,N \tag{10}$$

$$a_n(\xi, t = 0) = \begin{cases} a_{n,i} \ , & \text{for } \xi < \xi_i \\ a_{n,i+1}\ , & \text{for } \xi > \xi_i \end{cases} \tag{11}$$

The Riemann problem is a classical problem in the theory of hyperbolic equations (Lax (1973)). Its initial condition consists of two constant states to the left and right respectively of a discontinuity, which is located here at position ξ_i. The solution of the Riemann problem is formed by the two constant states separated by smooth segments (*rarefaction waves*) and discontinuities (*shocks*). The discontinuities must be constrained by the *Rankine-Hugoniot jump conditions* and *entropy conditions*, which ensure physical correctness of the shocks.

Riemann solvers form the basic building-blocks of our numerical method for solving the system (7–8). A large time step can be maintained only if interactions between waves and shocks belonging to *different* Riemann problems are fully taken into account. Such large-time-step methods have been proposed by Holden et al. (1988) and Leveque (1982). Smooth waves are here represented by a sequence of (small) discontinuities. Whenever two or more discontinuities coalesce, they are merged into one (new) discontinuity for which the Riemann problem is again solved. Since the entire algorithm operates only on discontinuities, there is (almost) no numerical diffusion.

It remains to describe the solution procedure for individual Riemann problems. For the case of *two-phase flow* (e.g., water and oil), equation (10–11) reduces to a *scalar* hyperbolic problem:

$$\frac{\partial s}{\partial t} + \frac{\partial}{\partial \xi} f(s) = q(s) \tag{12}$$

subject to the initial condition:

$$s(\xi, t = 0) = \begin{cases} s_i & , \quad \text{for } \xi < \xi_i \\ s_{i+1} & , \quad \text{for } \xi > \xi_i \end{cases} \tag{13}$$

Here s $(0 \leq s \leq 1)$ denotes the *saturation* of one of the phases (e.g., water). Consequently, $1 - s$ is the saturation of the other phase (e.g., oil). In two-phase flow problems, the water *fractional flow function* $f(s)$ usually has an *S-shape*, i.e., it satisfies the conditions:

- $f(0) = 0 \quad , \quad f(1) = 1$;
- $f'(0) = f'(1) = 0$;
- $f(s)$ has precisely one inflection point s_I, i.e., $f'' > 0$ for $0 \leq s < s_I$ and $f'' < 0$ for $s_I < s \leq 1$.

The following simple geometric principle for solving the Riemann problem (12–13) can be found in Concus and Proskurowski (1979):

- If $s_i > s_{i+1}$, then construct the *convex* envelope $h(s)$ of $f(s)$ over the interval $[s_{i+1}, s_i]$;
- If $s_i < s_{i+1}$, then construct the *concave* envelope $h(s)$ of $f(s)$ over the interval $[s_i, s_{i+1}]$.

In those areas, where $f(s) = h(s)$ the solution of (12–13) is a rarefaction wave; in areas with $f(s) \neq h(s)$ the solution is a shock. These techniques have been incorporated in a computer code capable of solving quite general two-phase flow problems in up to three space dimensions.

For *three-phase flow* (involving water, oil and gas) the situation is more complicated. Here the system (7–8) consists of two nonlinear, coupled equations:

$$\frac{\partial s_1}{\partial t} + \frac{\partial}{\partial \xi} f_1(s_1, s_2) = q_1(s_1, s_2) \tag{14}$$

$$\frac{\partial s_2}{\partial t} + \frac{\partial}{\partial \xi} f_2(s_1, s_2) = q_2(s_1, s_2) \tag{15}$$

As unknowns one could select the *saturation* of water (s_1) and gas (s_2). The oil saturation is then defined by $s_3 = 1 - s_1 - s_2$. For certain three-phase flow models however the *Jacobian matrix*:

$$\nabla_{s_1,s_2}\mathbf{f}(s_1,s_2) = \begin{pmatrix} \partial f_1/\partial s_1 & \partial f_1/\partial s_2 \\ \partial f_2/\partial s_1 & \partial f_2/\partial s_2 \end{pmatrix} \tag{16}$$

may have *complex* eigenvalues in some subregions of the saturation space. Bell, Trangenstein and Shubin (1986) have discussed such a model, which is formally *elliptic* rather than hyperbolic. The solution of Riemann problems associated with these equations of *mixed* type is very complicated. Also the method of characteristics can no longer be applied. Many aspects of these *elliptic regions* are not understood. Questions have been raised by many authors concerning the "correctness" of such models for three-phase flow (for example, see Fitt (1990)). Effective solution methods for the Riemann problem of three-phase flow are still under investigation.

4 Gridding

The computational scheme must be able to resolve accurately the moving fronts occuring in porous media flow. We adopt dynamic regridding, which must be computationally inexpensive. Clear coding of the mixed finite element discretisation is obtained when the grid consists of *blocks* on which the scalar quantities are stored, separated by *faces* on which the fluxes are stored. We also want *one* formulation with a dimension N_d that can be chosen as 1, 2, or 3. These considerations led us to the following *hierarchical* grid structure:

- A static regular *base grid* is created by orthogonal "planes" (constant value of a space co-ordinate). The rectangular boxes are called *blocks*. These are separated by rectangular "planes", called *faces*.
- The base grid is refined and unrefined dynamically by the *basic refinement*, which divides a block into 2^{N_d} identical smaller ones.

Our current codes have cartesian co-ordinate systems, but the theory covers also mildly (i.e., without singularities) curvi-linear co-ordinates. For curvi-linear co-ordinates with singularities, such as polar co-ordinates, additional theory is needed at least for the multigrid.

In the examples given in the next section, the grid is refined statically around wells. The dynamic refinement is triggered primarily by saturation. Given a saturation distribution, we refine and coarsen the grid to achieve

$$tol_s/4 < \epsilon_i < tol_s\,, \tag{17}$$

for all blocks i. Here ϵ_i is an indicator for the error in the saturation on block i based on approximations of first- and second-order space-derivatives of the saturation. The tolerance tol_s has a user-specified value. Each time step, the grid is refined iteratively, using saturation values both on the old and new time level. In each iteration a refinement (to just one level higher) is done if the upper bound in (17) is violated. The number of iterations does not exceed a prescribed maximum number of refinement levels. After refinement, the grid is coarsened in regions where the lower bound in (17) is violated, now using only saturations at the new time level. We can trigger on the total flow field $\mathbf{u}$ also, by introducing appropriate terms in the error indicator ϵ_i.

5 Numerical results

We present a number of numerical examples for two-phase flow, illustrating various aspects of the method. Fluids and rock formation are assumed to be incompressible. In all examples, the absolute permeability of the porous medium is constant. The first is a multi-well problem, the second a convergence study on the size of the time step, and the last a simple three-dimensional example.

5.1 Multi-well problem

The multi-well problem consists of three injection wells and one production well. The dimensionless size of the domain is 3×3. The injectors are located at $(0.9, 0.6)$, $(0.2, 2.8)$, and $(1.8, 2.1)$, with strengths 0.5, 0.2, and 0.3, respectively. Production occurs at $(3.0, 3.0)$. For the mobilities in (2) we take $\lambda_w = s^2$, $\lambda_o = (1-s)^2/2$. Figure 1 shows the solution at 0.67 PVI (*Pore Volume Injected*), after 12 time steps. The result clearly illustrates that numerical diffusion is virtually absent. Such a small amount of cross-flow smearing is difficult to obtain with standard finite difference methods.

5.2 Convergence

Our method should convergence when both the block size and the time step go to zero. Because the grid is adapted dynamically on the basis of a user-specified parameter, we only will consider convergence in terms of a decreasing time step.

The following set of examples is based on a simple inflow/outflow problem. The initial water saturation on the domain $\Omega = [0,1] \times [0,0.5]$ is defined as a hyperbolic tangent in the horizontal direction, which is perturbed by a cosine in the vertical direction. The inflow velocity is specified at the left boundary. At the right outflow boundary, the pressure is prescribed. The upper and lower boundaries are solid walls.

We have mobilities $\lambda_w = s^2$, $\lambda_o = (1-s)^2/2$, $\lambda = \lambda_w + \lambda_o$, and a fractional flow function $f(s) = \lambda_w/\lambda$. The endpoint mobility ratio is $\lambda(1)/\lambda(0) = 2$, which suggests instable displacement. The mobility ratio across the shock however is $\lambda(s_{shock})/\lambda(0) = 0.845$, which implies stable displacement. The latter mobility ratio seems to dominate the stability of the whole process.

Figures 2a–c show the saturation at the same time, but after different numbers of time steps. The results suggest convergence for decreasing time step, and also show that an acceptable solution can be obtained with relatively few time steps.

We have repeated this exercise for unstable displacement, using different mobilities that result in ratios $\lambda(s_{shock})/\lambda(0) = 2.051$ and $\lambda(1)/\lambda(0) = 7.262$. Figures 3a–c show the solution for an increasing number of time steps. Clearly, no convergence is obtained. The viscous fingering is generated by numerical "noise", which is amplified by the instability of the displacement process. The computations do not converge for $\Delta t \rightarrow 0$, which is in accordance with the ill-posedness of the problem.

A well-posed problem is obtained by adding a small amount of capillary pressure, which acts as a diffusion term and suppresses the instability on the smaller scales. Results are shown in Figs. 4a–c.

These experiments demonstrate that large time steps can be taken if the physical time scale of the problem allows this (no strong dependence of the total flow field on saturation).

5.3 A three-dimensional example

Figure 5 shows a solution of the three-dimensional equivalent of the quarter-five-spot problem. The injection- and production-wells are placed at opposite corners of a unit cube. The result has been obtained at 0.15 PVI in only one time step.

6 Conclusions

- In this paper, a numerical method is presented for computing multi-phase, miscible and immiscible flow through porous media in one, two, and three space dimensions.
- The method uses *adaptive local grid refinement* near wells, geological heterogeneities, and saturation fronts to enhance solution accuracy and to suppress numerical diffusion.
- Symmetric, elliptic equations are discretised by *mixed finite elements* and solved by *multigrid*. This combination provides accurate scalar and vector quantities and is efficient for large-scale reservoir simulation.
- The *method of characteristics* is used to solve hyperbolic systems of equations. The method introduces almost no numerical diffusion. Riemann solvers facilitate the treatment of nonlinearities and permit the use of large time steps.

Acknowledgement

The authors are indebted to the management of Shell Internationale Research Maatschappij BV for permission to publish this paper.

Notation and list of symbols

N_d	dimension $= 1, 2, 3$
$\mathbf{x}$, t	space and time co-ordinates
i , j	number of block, resp. face
n	$= 1, \ldots, N$, (chemical) component
m	$= 1, \ldots, M$, fluid-phase (e.g., water, oil, gas)
a_n , $\mathbf{u}_n$, q_n	concentration, flux, and sourceterm of component n
p_m , $\mathbf{v}_m$, λ_m , ρ_m	pressure, flux, relative mobility and density of phase m
$\mathbf{g}$	acceleration due to gravity
$\mathcal{K}$	(absolute) permeability tensor of porous medium
$\mathbf{u}$	total fluid velocity
ξ	parameter along streamline
s	(water) saturation
$f(s)$	fractional flow function
s_I	inflection point

References

Allen, M. B., Behie, G. A., Trangenstein, J. A.: Multiphase Flow in Porous Media. Springer Verlag (1988).

Bell, J. B., Trangenstein, J. A., Shubin, G. R.: Conservation laws of mixed type describing three-phase flow in porous media. SIAM J. Appl. Math. **46** (1986) 1000–1017.

Chavent, G., Cohen, G., Jaffre, J.: Discontinuous upwinding and mixed finite elements for two-phase flows in reservoir simulation. Computer Methods in Appl. Mechanics and Eng. **47** (1984) 93–118.

Concus, P., Proskurowski, W.: Numerical solution of a nonlinear hyperbolic equation by the random choice method. J. Comput. Phys. **30** (1979) 153–166.

Douglas, J., Yirang, Y.: Numerical simulation of immiscible flow in porous media based on combining the method of characteristics with mixed finite element procedures. In *Numerical Simulation in Oil Recovery*, M.F. Wheeler (ed.), Springer Verlag (1988) 119–131.

Ewing, R. E.: Problems arising in the modeling of processes for hydrocarbon recovery. In *The Mathematics of Reservoir Simulation*, R.E. Ewing (ed.), SIAM, Philadelphia (1983) 3–34.

Ewing, R. E.: Adaptive grid-refinement techniques for treating singularities, heterogeneities and dispersion. In *Numerical Simulation in Oil Recovery*, M.F. Wheeler (ed.), Springer Verlag (1988) 133–149.

Fitt, A. D.: Mixed hyperbolic-elliptic systems in industrial problems. In *Proc. Third European Conf. on Mathematics in Industry*, J. Manley et al. (eds.), Kluwer/Teubner, Stuttgart (1990) 205–214.

Gmelig Meyling, R. H. J.: Numerical methods for solving the nonlinear hyperbolic equations of porous media flow. In *Proceedings of the Third International Conference on Hyperbolic Problems*, Uppsala, June 1990 (to appear).

Holden, H., Holden, L., Hoegh-Krohn, R.: A numerical method for first order nonlinear scalar conservation laws in one dimension. Comp. Math. Appl. **15** (1988) 595–602.

Lax, P. D.: Hyperbolic systems of conservation laws and the mathematical theory of shock waves. Conf. Series in Applied Mathematics **11**, SIAM, Philadelphia (1973).

Leveque, R. J.: Large time step shock-capturing techniques for scalar conservation laws. SIAM J. Numer. Anal. **19** (1982) 1091–1109.

Miller, K.: Recent results on finite element methods with moving nodes. In *Accuracy Estimates and Adaptive Refinements in Finite Element Computations*, I. Babuska et al. (eds.), John Wiley (1986) 325–338.

Raviart, P. A., Thomas, J. M.: A mixed finite element method for second order elliptic problems. In *Mathematical Aspects of the Finite Element Methods*, Lecture Notes in Mathematics Vol. 606 Springer-Verlag (1977) 292–302.

Russell, T. F., Wheeler, M. F.: Finite element and finite difference methods for continuous flows in porous media. In *The Mathematics of Reservoir Simulation*, R.E. Ewing (ed.) (1983) 35–106.

Schmidt, G. H., Jacobs, F. J.: Adaptive local grid refinement and multigrid in numerical reservoir simulation. J. Comput. Phys. **77** (1988) 140–165.

Schmidt, G. H.: A dynamic grid generator and multigrid for numerical fluid dynamics. In *Proceedings of the* 8th *GAMM Conference on Numerical Methods in Fluid Mechanics*, P. Wesseling (ed.), Notes in Numerical Fluid Mechanics Vol. 29, Vieweg, Braunschweig (1990) 493–502.

Wesseling, P.: Multigrid methods in computational fluid dynamics. ZAMM **70** (1990) 337–347.

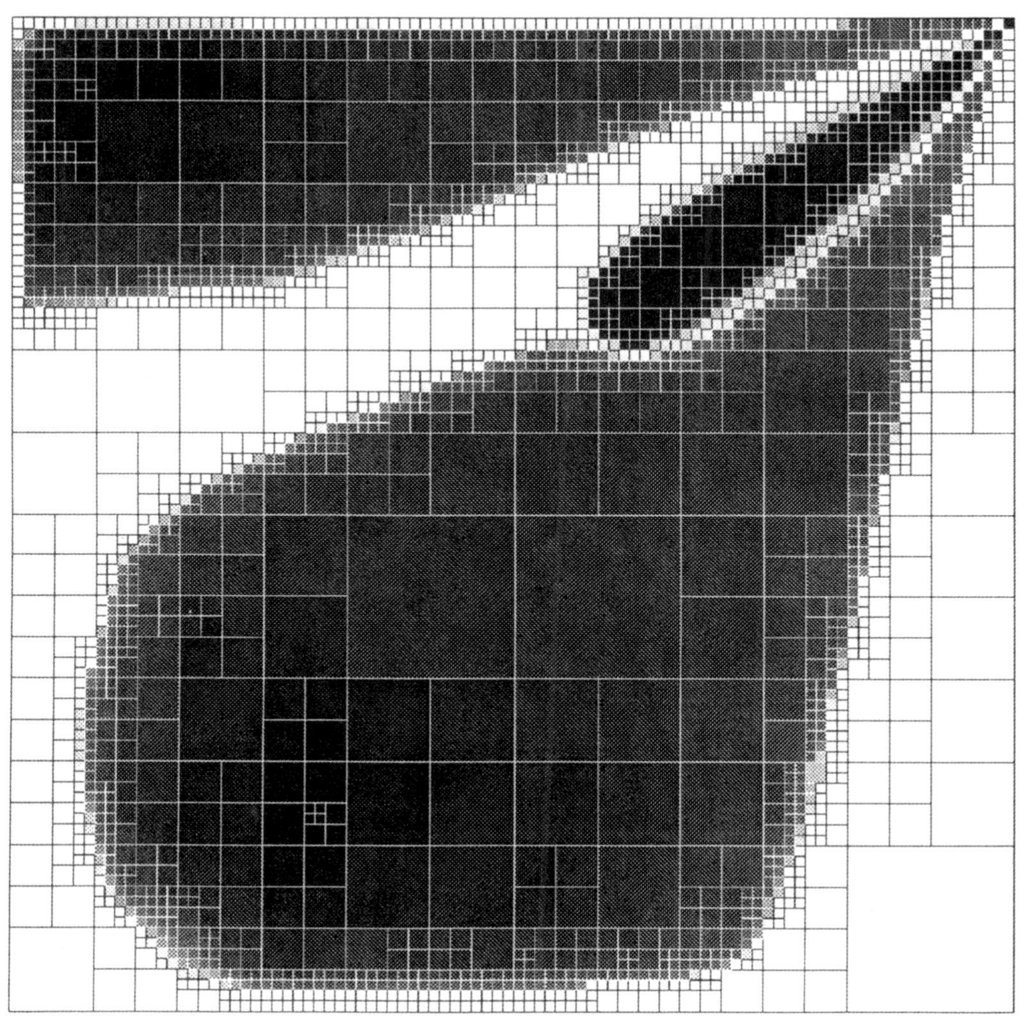

Fig. 1. Solution of the multi-well problem at 0.67 PVI using 12 time steps. Darker halftones correspond to higher water saturations.

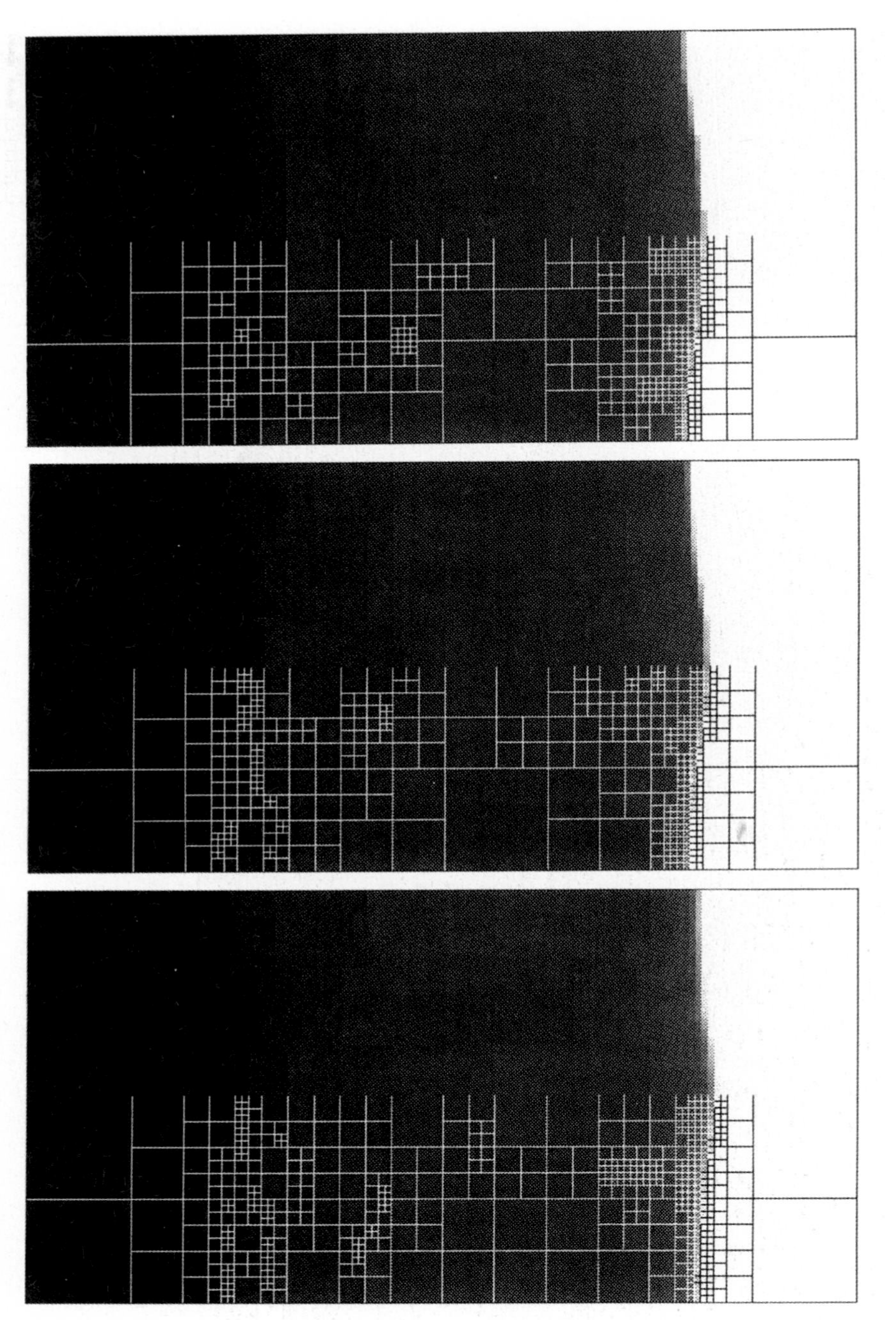

Fig. 2. Solution of the inflow/outflow problem for a stable displacement process with no capillary pressure using, from top to bottom, 32, 64, and 128 time steps.

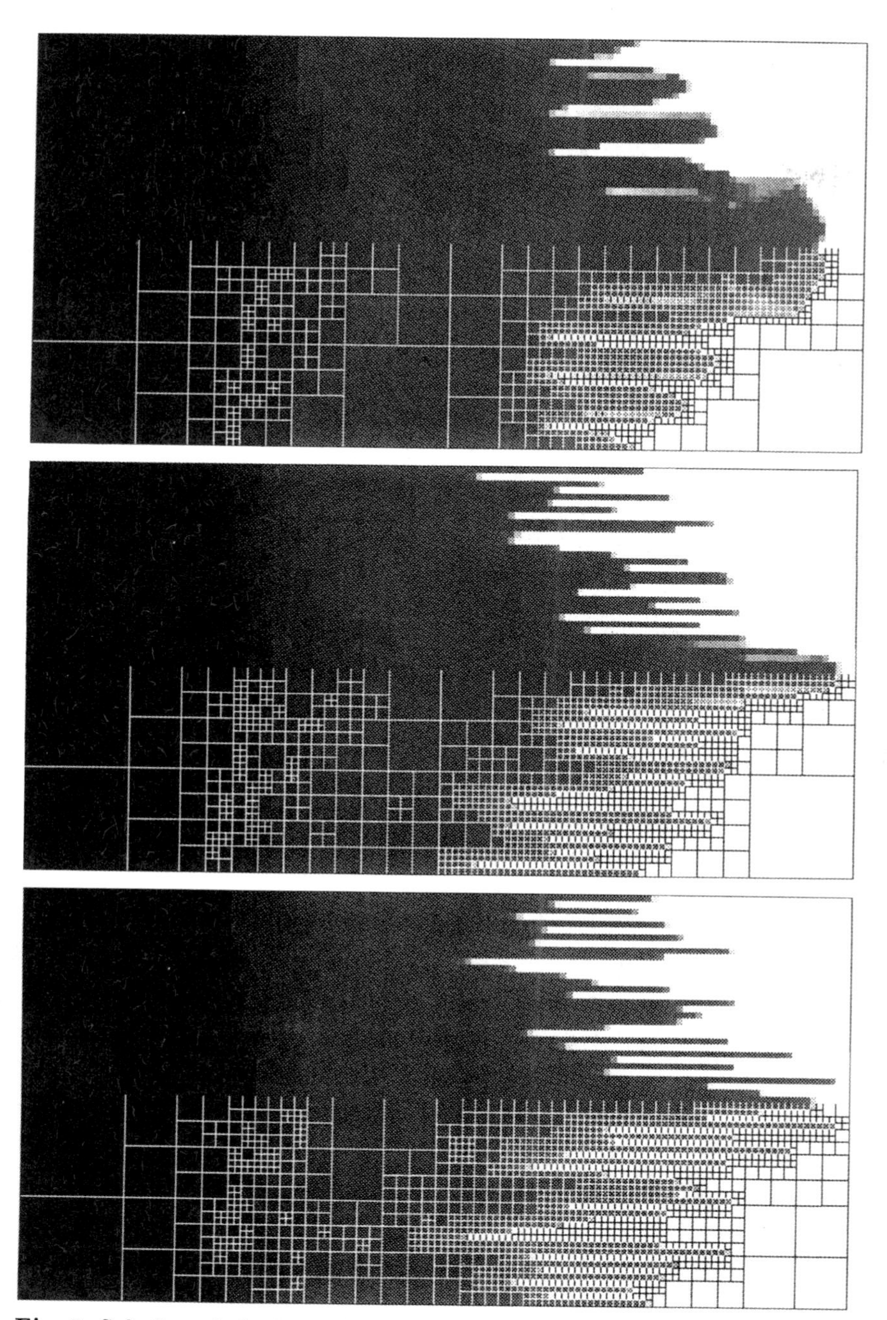

Fig. 3. Solution of the inflow/outflow problem for an unstable displacement process with no capillary pressure using, from top to bottom, 32, 64, and 128 time steps.

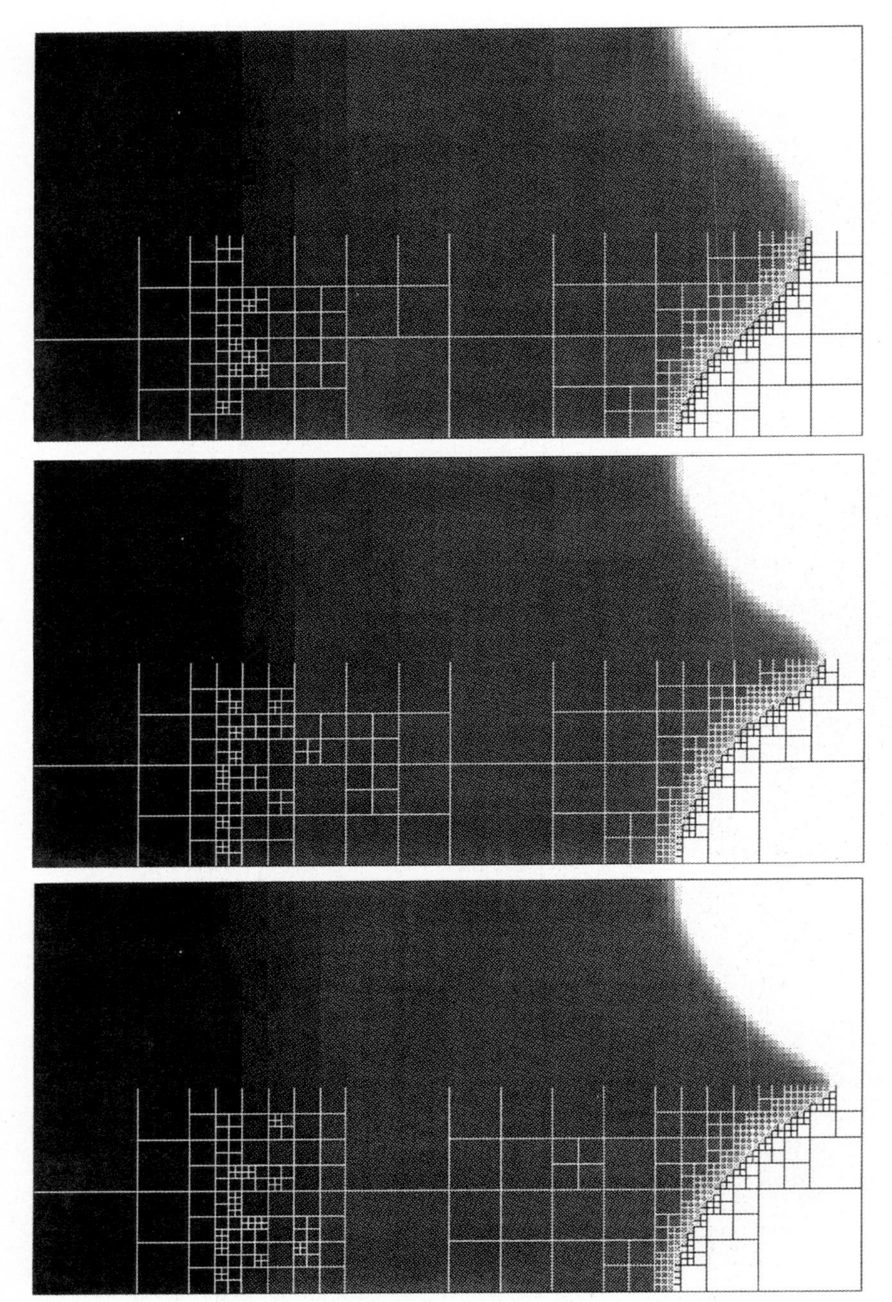

Fig. 4. Solution of the inflow/outflow problem for an unstable displacement process including capillary pressure using, from top to bottom, 32, 64, and 128 time steps.

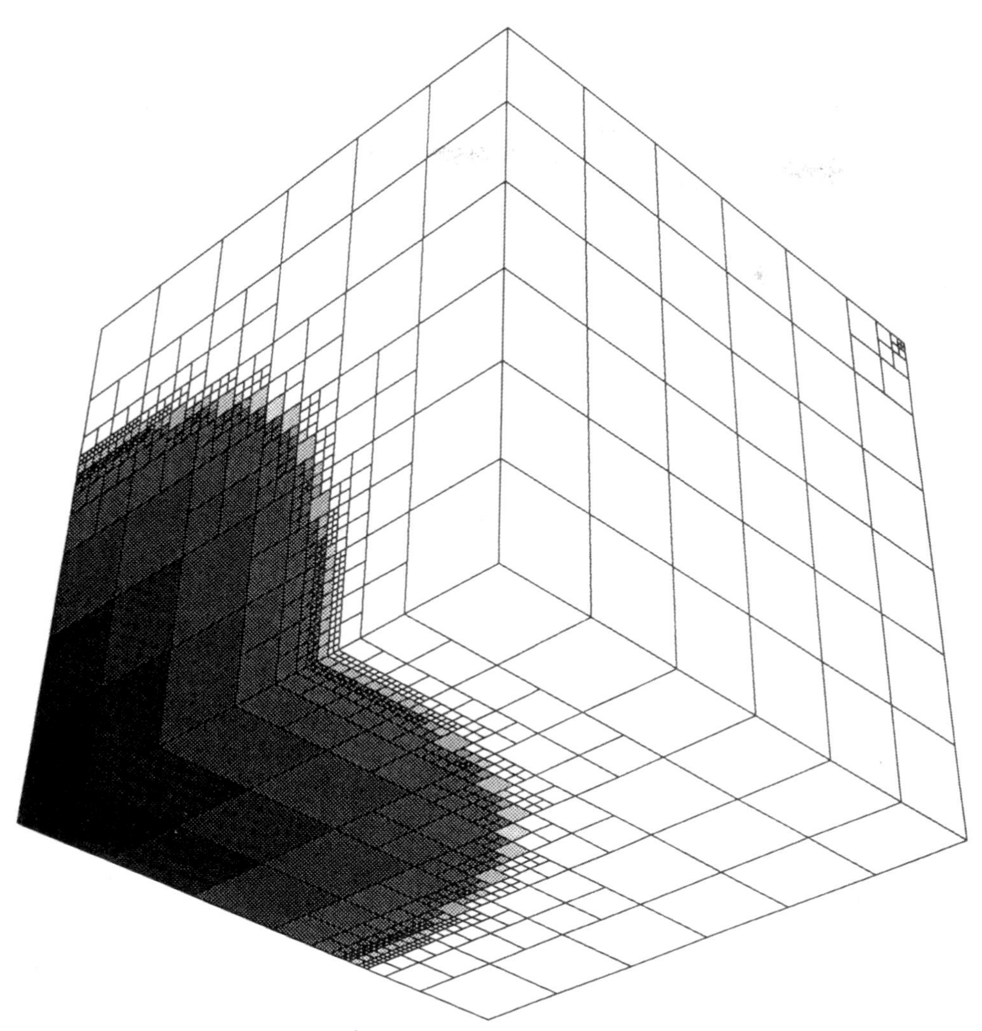

Fig. 5. Solution of the three-dimensional equivalent of the quarter-five-spot problem at 0.15 PVI using only one time step.

This article was processed using the LaTeX macro package with ICM style

FLOW AND DIFFUSION IN RANDOM POROUS MEDIA FROM LATTICE GAS SIMULATIONS

-=-=-=-=-

D. JEULIN

Centre de Géostatistique
E.N.S.M.P. 35 Rue Saint-Honoré 77305 FONTAINEBLEAU (France)

ABSTRACT

To study the effect of pore geometry on the overall permeability and diffusion properties of porous media, 2D lattice gas simulations were performed on an image analyzer :

. Using the F.H.P. model with six velocities on the hexagonal grid, and appropriate boundary conditions, flows in samples with infinite length are simulated, reproducing Poiseuille's flow conditions. Similar lattice gas experiments produced on random porous media simulations enable us to recover Darcy's law on a macroscopic scale. A systematic study of Darcy's permeability K was made on various simulations of the Boolean Random Set model. The following effects of pore geometry were investigated : statistical dispersion of K, relationship between K and the pore fraction, or the grain size, permeability of anisotropic porous media.

. With the same lattice gas simulations, dispersion properties of a porous medium are accessed : considering each coordinate of a tagged particle in the lattice gas as a diffusion stochastic process, second order statistics enable us to recover the macroscopic coefficient of diffusion of the porous medium, and to check the validity of a macroscopic Fick's law.

1. INTRODUCTION

The problem of the change of scale for flows in porous media is of primary importance to predict the effects of the pore geometry on the overall permeability of a porous medium. Among the well-known methods for estimating the permeability, we can mention the following :

o Using a probabilistic approach, G. Matheron solved the following problems :

- *Emergence of Darcy's law* on a macroscopic scale, from the linearized Navier-Stokes equations ruling the fluid flow on a microscopic level [3,6,7,8]. Concerning the incidence of pore geometry on the macroscopic permeability, upper bounds involving the second order moment of the pore chord size distribution were derived. No general and accurate estimator of the permeability is known. In fact the main difficulty is not to establish the validity of a macroscopic Darcy's law, but to account for the complexity of the boundary conditions to be respected by the flow through the complex geometry of the porous medium.

- *Composition of permeability* from one scale of measurement to a larger one : by variational techniques, bounds of permeability for infinite media are derived [2,4,5,6,8]. They involve a series expansion with correlation functions of the random permeability tensor with increasing order.

Two alternative techniques can be used to solve the composition of permeability problems. They are based on simulations :

- one can derive a random walk problem from a permeability map [1], and therefore estimate the macroscopic permeability from Monte-Carlo simulations on realizations of random permeability fields.

- one can calculate the solution of the flow equations using variational methods and Fourier series expansions on random field simulations [9,10].

o In the present text, we develop a different approach to recover Darcy's law from a microscopic flow in a porous medium [11,12,13]. It is based on so-called lattice gas simulations : a discretization of the space, time, and the velocity field is used, so that the medium and the fluid reside on a lattice. The fluid is made of discrete particles obeying collision rules. The strength of this method is that it operates at an "undermicroscopic" scale, where basic physical conservation rules are applied. At that scale, there is no partial differential equation to solve, and the boundary conditions are easy to handle. The simulation leaves a population of particles evolve, like a dynamic system, until a possible statistical equilibrium is reached. This is a typical simulation of a statistical physics problem.

After a description of the mode of simulation we have used, we will present examples of 2-D simulations of Darcy's law and of Fick's law on various random porous media.

2. LATTICE GAS SIMULATIONS OF FLUID FLOW

2.1. INTRODUCTION

The basic idea of lattice gas simulations is to consider a fluid as a population of moving particles on a lattice. It goes back at least to Broadwell [14], and was recently considerably extended to solve hydrodynamics problems, mainly in the field of turbulence [15,19].

2.2. BASIC RULES

We use the F.H.P. model developed by Frisch, Hasslacher [16] on a hexagonal lattice in two dimensions : we consider a gas of particles having the *six unit velocities* defined on the hexagonal grid (U_0 to U_5). At each point of the lattice, there is at most one particle with a given velocity. Therefore, the gas can be described by a set of six binary images (one per direction of its velocity). Each particle of unit mass moves on the lattice. In a unit time step, it will reach the nearest site of its initial position in the direction of its velocity. For the population of particles in one binary image it results in a *translation* of the image in the appropriate direction. In addition to the translation of the particles, we need rules of interaction between particles, namely *collision rules.* It is easy to check that the F.H.P. rules below involving head-on collisions and triple collisions preserve the overall momentum of the gas, and its overall mass. (For head-on collisions each issue is alternatively selected according to the parity of the time step ; similar results would be obtained from a random selection at each pixel, which would be a more appropriate rule of collisions, but would require the use of an additional bit plane with refreshed random values at each time step).

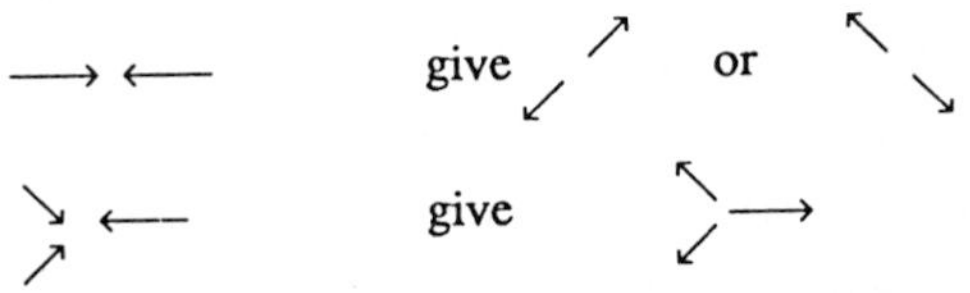

$\longrightarrow \longleftarrow$ give $\swarrow\!\nearrow$ or $\nwarrow\!\searrow$

$\searrow\nearrow \longleftarrow$ give $\nwarrow\swarrow \longrightarrow$

Therefore one cycle of the gas evolution is made of two steps :

- the translation of the six binary images,
- the changes in each binary image of the pixels involved in collisions to respect the rules.

Each cycle is repeated indefinitely, simulating the particles movements and interactions.

2.3. BOUNDARY CONDITIONS

Working on an infinite medium would need no other rules than the basic ones. However, we will work on finite fields. Furthermore, we want to simulate flows in porous channels. So we need to elaborate boundary conditions for the particles colliding with obstacles. We used two types of rules :

- particles arriving on *grain boundaries* or on the two *lateral edges* of the field are back-reflected. This reproduces the usual "non-slip" conditions for fluid flows, as the resulting average velocity is zero on the boundary. In the present simulations, the field is considered as a channel closed on its lateral sides, reproducing permeability experiments on rock samples.

- the right and left edges of the field are left open, so that particles can enter or exit (Fig. 1), as if the field were immersed in an infinite pipe. To respect the mass preservation, it is necessary to inject on each time step particles on both edges (left and right) with a uniform velocity distribution, and a uniform ordinate (the rate of injection n was empirically set up for a given mass density). Other authors inject particles on the left side only ("wind tunnel" conditions in [23], [12]). From our experience, this is not appropriate, since in the case of a porous medium, large regions in the shadow of grains may become empty of fluid during the process !!

To generate a pressure gradient in the channel, we introduce a bias in the velocity of injected particles, by means of a probability factor p, concerning the horizontal velocity component on the particles introduced on the left side. Increasing p from 0 to 1 increases the pressure gradient which develops a gradient of the density of particles. So at each time step, n $\left(\frac{1-p}{6}\right)$ particles with velocity u_2, u_3 or u_4 are introduced on the right side ; n $\left(\frac{1-p}{6}\right)$ particles with velocity u_1, u_0 or u_5 are introduced on the left side ; additional pn particles with velocity u_0 are introduced on the left side.

$$
\begin{array}{lccccl}
 & \nearrow & u_1 & u_2 & \nwarrow & \\
\frac{n(1-p)}{6} + pn & \longrightarrow & u_0 & u_3 & \longleftarrow & \frac{n(1-p)}{6} \\
 & \searrow & u_5 & u_4 & \swarrow &
\end{array}
$$

Fig. 1 : Boundary conditions on the right and left edges of the field

A similar procedure to force the flow is used by Kadanoff et al. [24], but this is obtained by changing certain velocity components at randomly (uniformly) chosen pixels in the field. This has the same effect as replacing the pressure gradient by a gravity field, with the advantage of keeping the average local density of particles constant. These simulations enable the determination of the permeability of a porous medium, with a forcing term replacing the pressure gradient.

Most authors use periodic, or quasi periodic conditions [12,23,24]. Others do not give any details on this point. Periodic conditions have the advantage of immersing the field into an infinite medium, and of letting it evolve without any perturbation of the boundary (in our procedure, we impose a uniform location of injected particles, which is not in agreement with a free evolution of the fluid). But periodic conditions are not the correct way to handle flows in a porous medium, since there is no reason to have in general any periodicity of the intersection between the edges of the field and the medium. An alternative way to solve this problem, suggested by G. Matheron, is first to reproduce the porous field into four adjacent fields by symmetry, and then to use periodic conditions. This will be used to measure the dispersion properties of a fluid flow in a porous medium in part 4.

2.4. IMPLEMENTATION AND INTERPRETATION OF THE SIMULATIONS

Each iteration of the process produces six binary maps for the velocities. It is easily implemented and at a low cost on a standard image analyzer, the Morphopericolor system operating on a hexagonal lattice. We can record, as a function of the iterative step, local or global data to characterize the process. We usually check the evolution of the total number of particles N(t) (to control the global mass preservation) and the horizontal average velocity component $\bar{u}_x(t)$, starting from a uniform random distribution. In all our simulations the average component reaches a nearly constant value after about 500 iterations where a statistical equilibrium of the velocity field was obtained. We can then study the property of this velocity field (Euler point of view), considered as a solution of the boundary value problem (Note that no partial derivative equations system was introduced).

In fact, from statistical physics considerations, it can be shown that the lattice gas rules (as those used in this report, and others) enable to recover the Navier-Stokes equations, when averaging the fluid flow on regions of appropriate size [13-19].

Instead of working on binary images, some authors suggested working on local probabilities ($0 \leq n_i \leq 1$), and applying the Boltzmann independence assumption at each pixel for the collision rule [L.B.E. model in 20,21,22]. The resulting simulations give velocity maps without noise, even for small systems. However, there is no guarantee that probabilities will be obtained in the calculation.

We can either mark a given particle, and follow its trajectory in time (Lagrangian point of view), or study the diffusive properties of the porous medium. The first point of view leads us to Darcy's law and the permeability estimation of the porous medium, as illustrated in part 3.

The second point of view will give us, at a macroscopic scale, the Fick's law of diffusion, and an estimation of the effective coefficient of diffusion of the porous medium, as illustrated in part 4.

3. PERMEABILITY OF POROUS MEDIA

This section is devoted to the results of simulations of flow in 2-D porous media from the lattice gas model. Starting from the microscopic model, and from binary images of porous media, we are looking for a change of scale, giving at the "macroscopic" level relationship between the flux and a pressure gradient. We will thus recover Darcy's law and estimate a permeability coefficient.

3.1. POISEUILLE'S FLOW IN A CHANNEL

It is well known from elementary fluid dynamics that a pressure gradient applied in the direction of two parallel flats separated by the distance d on the Oy axis generates a velocity field with the following properties :

$$\begin{cases} u_x(y) = |\overrightarrow{\text{grad}}\, p| \; \frac{1}{2\mu} \left[\frac{d^2}{4} - y^2 \right] \\ u_y(x) = 0 \end{cases} \tag{3.1}$$

where μ is the viscosity of the incompressible fluid, u_x and u_y being the horizontal and vertical components of the velocity.

From relation (3.1), the average x component of the velocity $\bar{u}_x$ is given by

$$\bar{u}_x = | \overrightarrow{\text{grad}}\, p \,|\, d^2/12\mu \tag{3.2}$$

It is easy to check experimentally the velocity of Poiseuille's law (3.1.) from a velocity map. Furthermore from a flow simulation the viscosity μ can be obtained experimentally, as proposed by Kadanoff et al. [24].

The proportionality between the flux $\bar{u}_x$ and the pressure gradient in relation (3.2) can be interpreted as Darcy's law of the non porous medium, made of the channel with width d. With this interpretation the medium has a permeability $K_x = \frac{d^2}{12}$. However, this is not an intrinsic property of a medium, as it depends on a size parameter (d) of the sample. But, it may be of interest to compare the value K_x obtained on the non porous channel and on porous channels.

Poiseuille's flows were simulated by the lattice gas method according to the following conditions :

- the boundary conditions were applied as defined in section 2.3.
- different pressure gradients were used.

- 2000 iterations (20 min. on the Morphopericolor system) were performed on a population of 10^5 particles in a field of length 256 and width 200.

The results of the simulation allow the following comments :

- After a transition period (nearly 500 iterations), the average flux $\bar{u}_x$ remains nearly constant.

- The velocity profiles (derived from the velocity field obtained at iteration N°2000 by averaging the velocity on every line of the field) are in good agreement with the predicted parabolic profile (equation (3.1)).

The flux obtained for the maximum applied pressure gradient ($p = 1$ as defined in section 2.3.) is equal to 0.483.

The results obtained in this section can be considered as a test of the validity of the method of simulation : there is a good agreement between a theoretical and a simulated profile in the case of the well known Poiseuille's flow.

3.2. FLOW IN A POROUS MEDIUM

The same approach as in subsection 3.1. is now applied to the case of a porous medium (grains are introduced in the channel), where an increasing pressure gradient is used. As before, an almost stationary flux is reached after nearly 500 iterations (see fig. 2 below). We also get a proportional variation of the flux with the pressure gradient parameter p. It is an experimental verification of Darcy's law. The proportionality constant is related to the permeability K. As the remaining simulations are made in the same conditions (size of the channel, horizontal pressure gradient,...), except for the porous medium, we will consider this proportionality constant as a permeability (with appropriate units including the physical size of the sample, and the viscosity μ). In order to reduce their number, the remaining simulations were made with one pressure gradient ($p = 1$), and the permeability was estimated from the corresponding flux after 2000 iterations. Thus the permeability of one porous field was obtained from a 20 min simulation, which is quite satisfactory.

3.3. INFLUENCE OF SOME GEOMETRICAL PARAMETERS OF THE POROUS MEDIUM ON ITS PERMEABILITY

For this preliminary study, a limited range of geometrical parameters were investigated. Our porous media are in fact bounded domains of Boolean models [6] with square or rectangular grains. We considered the following points :

i) dispersion of the permeability induced by the spatial distribution of grains

ii) influence of the pore volume fraction on the permeability

iii) influence of the grain size on the permeability

iv) permeability of anisotropic media

i) Effect of the spatial grain distribution on the permeability

In all our simulations on porous media we operated on finite size fields. At this scale, geometrical fluctuations between different realizations of a random model of porous medium induce fluctuations of the permeability. This is illustrated by a simulation of ten fields for a uniform implantation of 80(10×10) square grains. In these simulations, the porosity is nearly constant ($\simeq$ 0.15), but the distribution of grains changes so that critical channels slowing down the flow have different positions and sizes in the realizations of the medium. A useful indication of the resulting variability is given by the statistical standard deviation σ of the experimental permeabilities (σ = 0.0181 for an average value $\overline{K}$ of 0.0971) and by the coefficient of variation $\sigma/\overline{K}$ equal to 0.1866. We can draw the following conclusions from these data :

- The average value of K over 10 simulations is known with 11.8% relative precision. However, a medium made of the arrangement of different fields built from the same model would not possess a global permeability equal to the arithmetic average of individual permeabilities. It is well known that there is generally no such composition law [2,4-6].

- At the scale used for the simulations, the medium is not spatially ergodic, from the point of view of its permeability (despite the quasi ergodic behavior of the pore fraction). Therefore, at this scale, the "macroscopic permeability" as obtained from flow simulation is a random variable. We will characterize it by a histogram of permeability.

Concerning the change of scale in the porous medium, this could be done in two different ways :

o using larger fields, one can expect to have an ergodic situation where K is not a random variable anymore, if there is not a full range of scales in the porous medium.

o a first change of scale might be operated by our flow simulations on the available images, giving a permeability K(x) (x being the center of the image). It could be followed by a second change of scale applied to the field K(x), using other methods, either probabilistic [2,4-6] or finite element or finite difference numerical techniques [9,10].

To investigate the influence of other geometrical properties of the porous medium on its permeability, the following types of random porous media simulations were performed :

- *Percolation medium model* (each pixel is independently assigned to grains with a given probability) for different grain proportions : 0.1, 0.15, 0.2. For each proportion 10 simulations are provided. This model is made of a random implantation of grains X' (according to a Poisson point process) with possible overlaps of the grains, resulting in an interconnected porous medium.

- *Boolean models* with square random grains. Two sizes of primary grains were used : squares of size 5×5 and squares of size 10×10. Ten simulations are made for each size and for the following grain area fractions : 0.1 ; 0.2 ; 0.3 ; 0.4.

- *Boolean models with rectangular primary grains* (their longest side being parallel to the 0x axis) give anisotropic porous media : 10×20 and 10×30 rectangles were used, giving a constant grain area fraction (nearly 0.1). Ten simulations were obtained in each case. "Horizontal" (x direction) and vertical (y direction) flows were simulated. The results are illustrated in Fig. 2. The velocity maps are obtained from the velocity field after 2000 iterations, by averaging it into hexagons with side 8, allowing us to make the flow lines clearly appear.

Each simulation of a Boolean porous medium field is obtained in three steps :

- We use Poisson germs x_i for the location of the random grains X'_i. These germs are located in a field Z', larger than the field Z used for flow simulations (in fact $Z' = Z \oplus X'$, where $\oplus$ is the dilation operation. In our case, we dilate a rectangle Z ($N_1 \times N_2$) with a square or a rectangle $X'(N_3 \times N_4)$ resulting in a rectangle with edges $(N_1 + N_3)$ and $(N_2 + N_4)$.

For each simulation of the porous medium, the number of grains follows a Poisson distribution with density θ, so that the average pore area fraction ϕ is obtained by

$$\phi = \exp - \theta\ A(X') \tag{2.3}$$

A(X') being the area of the individual grain X' used in the construction of the model.

- For a given number of grains, their coordinates are generated by independent uniform random numbers over $[0, N_1 + N_3]$ and $[0, N_2 + N_4]$, as a property of the Poisson point process.

- A grain is centered at each location of the Poisson point process generated in the first two steps of the simulation. Overlapping between grains is allowed.

As illustrated by Fig. 3, showing histograms of permeabilities, there is a dispersion of the results of the simulations, due to the variance of the

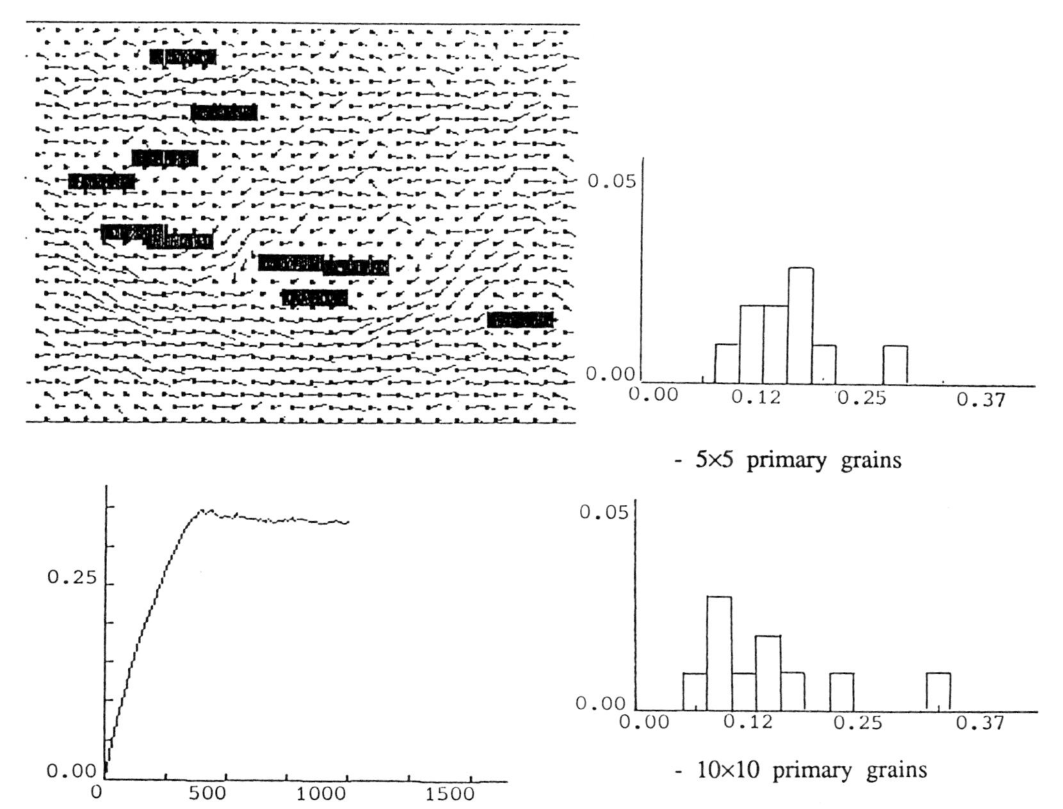

Figure 2 : Flow simulation in a 2-D porous medium (realization of a Boolean set with 10 × 30 "horizontal" primary grains) (top) Horizontal pressure gradient. Horizontal flux ($\overline{u}_x$) as a function of the number of iterations N of the lattice gas (bottom).

Figure 3 : Histogram of the experimental permeability K_{xx} (calculated from the sill of the curve ($\overline{u}_x$, N)) for realizations of boolean porous media.

Poisson point process and to the spatial distribution of the points. Again the size of the fields is too low to get an ergodic situation concerning the microstructure as well as the permeability. A different situation is obtained for the percolation model, where practically no dispersion occurs between realizations.

From this set of simulations, it is possible to draw some conclusions on the relationship between the permeability and various geometrical parameters, as explained below.

ii) Effect of the pore fraction on the permeability

For each case-study the permeability decreases with the grain fraction, as expected. Of course, there is no general relationship between a single scalar parameter (the pore fraction) and the permeability. In fact we looked for "local" relationships, valid on a given set of simulations, i.e. with a fixed primary grain (for the Boolean structures), or for the percolation medium.

It is clear from our results that the permeability reaches a null value before a 50% grain fraction. At this fraction, grain clusters may cross the image so that there is not any connected path for the fluid to flow. This is an effect of grain percolation. For this reason, no flow simulations were performed at higher grain fractions. This percolation effect, which appears at relatively low grain volume fraction in 2-D is certainly the major drawback when an estimation of the permeability of 3-D porous media is looked for. In the 3-D space, more possibilities do exist for the flow to avoid the grains. The only way to make proper simulations would be to work on 3-D models, using 3-D collision rules [19], or the L.B.E. model [22] at the expense of more complex and slower computations.

To compare the permeability-pore fraction relationship in the various simulations, it was convenient to look for an analytical expression. We found that an exponential law could be applied :

$$K = K_o \exp - \alpha\ (1 - \phi) \tag{3.4}$$

ϕ being the pore fraction of a given image, and K its permeability obtained from the flow simulation.

Rel.(3.4) has no theoretical foundation, and can be proposed on an empirical basis. In fact it gave a correct analytical representation of our data as seen on table 1, where the coefficients were estimated from a linear regression between the variables log K and ϕ.

Table 1 : Coefficients of regression for $K = K_0 \exp(-\alpha (1 - \phi))$

Model	Percolation *	Boolean Model					
		Squares 5×5 **	Squares 10×10	Rectangles 10×20 Horizontal flow	Rectangles 10×20 Vertical flow	Rectangles 10×30 Horizontal flow	Rectangles 10×30 Vertical flow
K_0	0.0275	0.2794	0.2367	0.453	0.5326	0.404	0.37
α	9.39	10.17	6.146	7.14	9.83	5.72	7.55
Correlation Coeff.	0.996	0.998	0.974	0.995	0.99	0.989	0.979

* three data - $1 - \phi \leq 0.2$
** simulation for $1 - \phi \leq 0.19$ available

The effect of the pore fraction on the decrease of the permeability can be explained from the coefficient α in relation (3.4), as detailed in the following subsections.

iii) Effect of the grain size on the permeability

A first effect can be given from table 1 : the coefficient α decreases with the grain size (compare 5×5 and 10×10 ; compare 10×20 and 10×30) : the grain fraction has less effect on decreasing permeability for larger grains (which are more distant from one another for a given grain fraction).

Another comparison can be made from table 2 where the variations of K with ϕ was rederived from relation (3.4) and the coefficients in table 1 for the 5×5 and 10×10 squares. In fact, for a given volume fraction, changing the size of the squares is equivalent to a homothetics of the microstructure, and it is known that increasing the size by a factor λ should increase the permeability by λ^2. From table 2, this is approximately verified for the grain fraction 0.35 (where the results are extrapolated in the 5×5 case). For lower grain fractions, it seems that the velocity profile is not far from the Poiseuille simulation in both cases, so that the apparent permeability is similar. This can be considered as an edge effect of our boundary conditions appearing for low grain fractions. Other boundary conditions, as mentioned in section 2, would be more appropriate for this case.

In addition, it is expected that increasing the grain size should increase K_0 (in table 1) by a factor λ^2, and should leave the coefficient α nearly constant. This is obviously not the case, as a result of the finite size of the simulated fields for which the spatial ergodicity is not obtained.

Table 2 : Relation between K and 1 - ϕ (from rel 2.5 and table 2.7) on Boolean structures

Solid fraction $1 - \phi$	Squares 5×5	Squares 10×10
0.1	0.101	0.128
0.15	0.061	0.094
0.2	0.036	0.069
0.25	* 0.022	0.0509
0.3	* 0.0132	0.0375
0.35	* 0.00795	0.0275
	* : extrapolation	

iv) Permeability of anisotropic porous media

From flows simulated in two orthogonal directions on anisotropic media (Fig. 2), information on the two diagonal components of the permeability tensor, K_{xx} and K_{yy}, is obtained. In the previous cases, nearly isotropic media had an isotropic tensor of permeability, that could be reduced to a scalar K (in fact with square primary grains we should have $K_{xx} = K_{yy}$ in our system of coordinates, and a slight off-diagonal component K_{xy}).

A global effect of the anisotropy is the following (see table 1) : the coefficient α increases from the horizontal flow to the vertical flow, so that the grain fraction alone cannot explain the change of permeability for anisotropic media, as expected. An increase of the grain fraction has a larger effect on the vertical flow (the horizontal grains acting as barriers) than on the horizontal flow. A further comparison is made from table 3, derived from relation (3.4) and the appropriate coefficients in table 1 :

Table 3 : Relation between K and $1 - \phi$ on anisotropic Boolean structures.

Solid fraction $1 - \phi$	Rectangles 10×20 Horizontal flow K_{xx}	Rectangles 10×20 Vertical flow K_{yy}	Rectangles 10×30 Horizontal flow K_{xx}	Rectangles 10×30 Vertical flow K_{yy}
0.1	0.222	0.199	0.228	0.174
0.15	0.155	0.122	0.171	0.119
0.2	0.109	0.074	0.129	0.082
0.25	0.076	0.046	0.097	0.056
0.3	0.053	0.028	0.073	0.038
0.35	0.037	0.017	0.054	0.026
0.4	0.026	0.010	0.041	0.018
0.45	0.018	0.006	0.031	0.012

At low grain fraction the effect of the grain size is not important, but we still have $K_{xx} > K_{yy}$ for $1 - \phi = 0.1$. The ratio K_{xx}/K_{yy} increases with $1 - \phi$ (as $\alpha_y > \alpha_x$) : increasing the number of grains increases the number of barriers to the vertical flow, which act as preferential channels for the horizontal flow. At low grain fractions, the ratio K_{xx}/K_{yy} is larger for larger grains. For $1 - \phi \geq 0.35$, these ratios are similar in the two studied cases (for the same grain fraction, the structures become similar, due to the grain overlaps).

For a given grain fraction (in the range $0 < 1 - \phi < 0.2$) K_{xx} and K_{yy} are always larger for anisotropic media than for the isotropic 10×10 square medium (table 2) : the former case involves larger channels. So it appears that the width of channels is of major importance in explaining the permeability.

A local anisotropy coefficient for the permeability can be obtained by the ratio K_{xx}/K_{yy} calculated for each field of simulation. On average this ratio is larger for 10×30 rectangles. However, the fluctuations are important, due to the large spectrum of grain fraction and to the spatial distribution of the grains.

4. DISPERSION IN POROUS MEDIA

4.1. PRINCIPLE OF DIFFUSION FROM LATTICE GAS SIMULATIONS

In this section, we introduce an approach of the diffusion in porous media based on random walks considerations. A different approach, based on the time evolution of concentrations, is proposed in [26].

Instead of considering the velocity field obtained in lattice gas simulations, it is possible to mark a given particle and to follow its trajectory with time. With this Lagrangian point of view for the fluid flow, we can study a given trajectory as a random walk (a diffusion process or a Brownian motion with an advective speed u when it is constant ; see Fig. 4 for a practical simulation). We can point out that in the present case, the random walk is just the result of the interaction between the marked particle and the other particles in the fluid, respecting the boundary conditions. In the simulation, the velocity of the particle after each collision is randomly selected amongst possible velocities. As in [25], we can consider the trajectory of a particle in a random velocity field U(x). Each coordinate $X_i(t)(i = 1,2)$ of the particle starting from $x(x_i)$ at $t = 0$ is a diffusion stochastic process with expectation and covariance given by

$$E[X_i(t)] = x_i + \bar{u}_i\, t \tag{4.1}$$

$$E[X_i(t) - x_i - \bar{u}_i\, t]\,[X_j(t) - x_j - \bar{u}_j\, t] = 2D_{ij}\, t \tag{4.2}$$

when a macroscopic Fick's law is observed.

In relation (4.1) $\bar{u}_i$ is the average i component of the velocity fluid, while in relation (4.2), D_{ij} is the effective matrix coefficient of diffusion. It should be noticed that the macroscopic coefficients $\bar{u}_i$ and D_{ij} obtained from expectations of various particle trajectories are valid for an equivalent homogeneous medium if for this medium the conditions for a macroscopic Fick's law are fulfilled. It is difficult to know these conditions in general. In [25] the advective random field u_i is assumed to be stationary. For a self similar (and therefore non stationary) porous network (4.1) and (4.2) are not valid, and a t^{α} behavior may be observed (anomalous diffusion). In practice, for our simulations, we checked the validity of relation (4.2) from the experimental variograms (with $X_1(t) = X(t)$ and $X_2(t) = Y(t)$) ;

$$2\gamma_x(\Delta t) = E\,[X(t + \Delta t) - X(t)]^2 = 2D_x\Delta t + u_x^2\,(\Delta t)^2 \tag{4.3}$$

$$2\gamma\,(\Delta t) = 2\gamma_x(\Delta t) + 2\gamma_y(\Delta t) = 2(D_x + D_y)\Delta t + (u_x^2 + u_y^2)\,(\Delta t)^2 \tag{4.4}$$

for which a fit to a parabolic curve was looked for.

Additional information might be obtained from the empirical histogram of the sojourn time τ of the particle in the field. If we consider the particles leaving the field at the abscissa a, we should expect, in the case of a constant velocity field $(\bar{u}_x, \bar{u}_y)$, and on an infinite medium :

$$F\,\tau_a(t) = P\{\tau_a \le t\} = \sqrt{\frac{1}{\Pi D_x t}} \int_a^{\infty} e^{-\left(\frac{(x - \bar{u}_x t)^2}{4 D_x t}\right)}\, dx \tag{4.5}$$

For a non-homogeneous medium, we would expect a similar distribution function involving the effective $\bar{u}_x$ and D_x of the equivalent homogeneous medium.

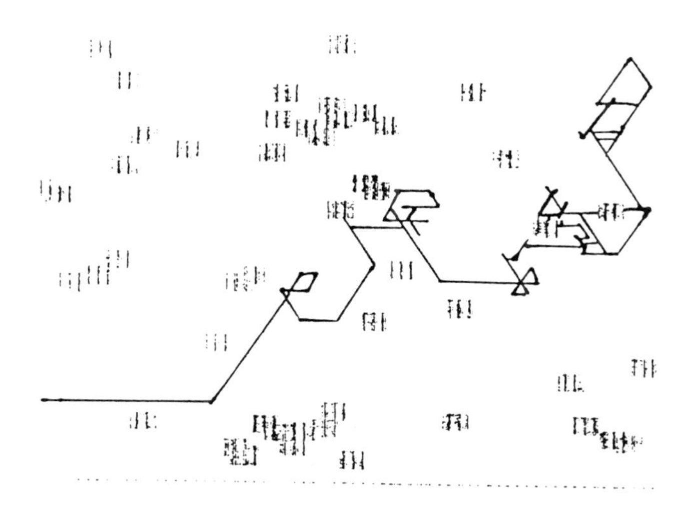

Figure 4 : Trajectory of a particle during a Poiseuille's flow simulation in a porous medium (Boolean model with 10×10 primary grains).

4.2. Dispersion in porous media : Preliminary results

The dispersion of particles was simulated in various experiments with an average of 1.9 particles per site. The preliminary results are the following [27] :

- Simulations of the *dispersion of particles in a gas* were performed under two different conditions :

o Six simulations on a 64x64 window under periodic boundary conditions (as explained in section 2.3) to reproduce an infinite medium, $(\bar{u}) = 0)$ for 3 600 time steps. In that case, $2\gamma(\Delta t)$ is a straight line (correlation coefficient 0.999) with a slope $2(D_x + D_y) \simeq 6.86$.

o An additional simulation on a Poiseuille's flow (under periodic conditions on the two ends of the pipe) forced by a gravity field $(|\bar{u}_x| = 0.25)$ gave very similar results $(D_x + D_y = 3.5$ with a correlation coefficient 0.98).

For this range of condition $(0 < |\bar{u}_x| \leq 0.25)$, and this particle density, we therefore can consider that $D_x + D_y \simeq 3.5$.

- Finally, *the dispersion in a porous medium* (Boolean model with 10×10 grains so that $\phi = 0.9$ on a 64×64 window followed by a symmetrization and periodic boundary conditions) for $|\bar{u}| = 0$ and 4800 time steps gave $2(D_x + D_y) = 4.95$ (correlation coefficient 0.9978). For this type of porous medium, the dispersion is slowed down by the presence of grains, and there is no anomalous diffusion, so that the macroscopic Fick's law (with an effective diffusion coefficient) is valid.

The preliminary study should be completed by a more systematic approach including the effect of various pore geometries, as well as of the hydrodynamic conditions, as D is expected to change with $|\bar{u}_x|$.

5. CONCLUSION

In this study the feasibility of lattice gas simulations to estimate the effective permeability and coefficient of diffusion was demonstrated.

Using a standard image analyzer operating on binary images of a porous medium was efficient to simulate a fluid flow, and provided results on the effect of pore fraction, of the grain size, and of anisotropy on the permeability K. The estimated K, which can even be non scalar, can be introduced into a composition of permeability procedure for investigating higher scales in porous media.

In addition some other topics are worth studying in depth :

- On an actual porous medium, a comparison between simulations and physical measurements of the permeability should be made. It should be of major interest to determine at which scale images of the porous medium should be used for simulations. Indications on the details of the microstructure which are crucial for the physical properties could be obtained from progressive simplifications on the medium by appropriate morphological filters, such as opening or closing transformations.

- To have a more realistic description of the effect of the pore connectivity, 3-D simulations should be developed.

- Relationships between the permeability and morphological data measured on the same fields (as was illustrated in our simulations) should be looked for, in order to infer the permeability or the diffusion coefficient from simple measurements, such as histograms of distance [6,7] or geodesic criteria [28,29].

ACKNOWLEDGEMENTS : The author is grateful to Tang Chang Qing and to R. Bremond for the implementation of the simulations on an image analyzer.

REFERENCES

[1] G. Matheron : "Equation de la chaleur, écoulements en milieux poreux et diffusion géochimique". (Oct. 1964), Paris School of Mines Publications.

[2] G. Matheron : "Structure et composition des perméabilités". Rev. Inst. Franç. du Pétrole, 1966, XXI,4, 564.

[3] G. Matheron : "Génèse et signification énergétiques de la loi de Darcy". Rev. Inst. Franç. du Pétrole, 1966, XXI,11, 1697.

[4] G. Matheron : "Composition des perméabilités en milieu poreux hétérogène: méthode de Schwydler et règles de pondération". Rev. Inst. Franç. du Pétrole, 1967, XXII,3, 443.

[5] G. Matheron : "Composition des perméabilités en milieu poreux hétérogène: critique de la règle de pondération géométrique". Rev. Inst. Franç. du Pétrole, 1968, XXIII,2, 201.

[6] G. Matheron : Eléments pour une Théorie des Milieux Poreux (Masson, Paris, 1967).

[7] G. Matheron : "L'émergence de la loi de Darcy". (1979) Paris School of Mines Publications.

[8] G. Matheron, Ann. Mines 5-6 (1984) 11.

[9] G. Le Loc'h-Lashermes, Thesis, Paris School of Mines (1987).

[10] G. Le Loc'h-Lashermes, in : Proc. 3rd Int. Geostatistics Congress, Avignon, France, 5-9 Sept. 1988 (Reidel, Dordrecht).

[11] D. Jeulin, Tang Chang Qing : "Lattice gas flow simulations o, 2-D random porous media". (1990) Paris School of Mines Publications.

[12] D.H. Rothman, Geophysics 53 (1988) 509.

[13] K. Balasubramanian, F. Hayot, W.F. Saam : "Darcy's law from lattice-gas hydrodynamics". Phys. Rev. A, **36**, 2248 (1987).

[14] J.E. Broadwell : "Study of rarefied shear flow by the discrete velocity method". J. Fluid Mech. 19, 401 (1964).

[15] J. Hardy, O. De Pazzis and Y. Pomeau, Phys. Rev. A 13 (1976) 1949.

[16] U. Frisch, B. Hasslacher and Y. Pomeau, Phys. Rev. Lett. 56 (1986) 1505.

[17] D. d'Humières and P. Lallemand, Physica A 140 (1986) 337.

[18] S. Wolfram, J. Stat. Phys. 45 (1986) 471.

[19] U. Frisch et al., Complex Systems, 1, (1987), 649.

[20] G.R. Mc Namara, G. Zanetti : "Use of the Boltzmann Equation to simulate lattice-gas automata". Phys. Rev. Lett., **61**, 2332 (1988).

[21] F.J. Higuera,J. Jimenez : "Boltzmann approach to lattice-gas simulations" Europhys. Lett., **9**, (7), 663-668, (1989).

[22] S. Succi, E. Foti, F. Higuera : "Three-dimensional flows in complex geometrics with the lattice Boltzmann method". Europhys. Lett., **10**, (5), 433-438, (1989).

[23] D. d'Humières, P. Lallemand : "Numerical simulations of hydrodynamics with lattice-gas automata in two dimensions". Complex Systems, 1, (1987), 539.

[24] L.P. Kadanoff, G.R. Mc Namara, G. Zanetti : "A Poiseuille viscometer for lattice gas automata". Complex Systems, 1, (1987), 791.

[25] G. Matheron : "Quelques exemples simples d'émergence d'un demi-groupe de dispersion". Internal Report (1979), Paris School of Mines.

[26] C. Baudet, J-P. Hulin, P. Lallemand, D. d'Humières : "Lattice-gas automata : a model for the simulation of dispersion phenomena". Phys. Fluids A, **1**, (3), March 1989, 507-512.

[27] R. Bremond : Paris School of Mines Publications, to appear in September 1990.

[28] L. Vincent, D. Jeulin : "Minimal paths and crack propagation simulations" Acta Stereologica, **8**, (2), 487-494 (1989).

[29] D. Jeulin, M.B. Kurdy : "Quelques paramètres géodésiques pour caractériser la connexité de milieux biphasés". (November 1989), Paris School of Mines Publications.

Lattice Gas Automata Simulations of Viscous Fingering in a Porous Medium.

J. F. Lutsko[1], *J. P. Boon*[1] *and J. A. Somers*[2]

[1] Faculté des Sciences, CP231
Université Libre de Bruxelles
B-1050 Bruxelles, Belgium
[2] Koninklijke/Shell-Laboratorium, Amsterdam (Shell Research B.V.)
Postbus 3003, 1003 AA Amsterdam, the Netherlands

1 Introduction

The study of viscous fingering began with the work of Saffman and Taylor and has been an active area of research ever since (for a review, see ref.[1]). Originally, it was of academic interest as an example of a hydrodynamic instability which, being essentially two dimensional, could be described by relatively simple approximations to the Navier-Stokes equations. The canonical experiment consists of a fluid with relatively high viscosity confined between two parallel glass plates separated by a narrow gap. A second fluid of relatively low viscosity is forced in at one end of this Hele-Shaw cell and displaces the first fluid which is allowed to flow out of the cell at the other end. If the interface between the fluids is originally planar, an instability develops which gives rise to the formation of fingers of less viscous fluid penetrating into the more viscous fluid. The physical problem consists of predicting the number and width of the fingers. Although considerable effort has gone into this problem, a complete theoretical understanding is still lacking.

A related phenomena occurs when the void between the plates is filled with a porous medium through which the liquids flow(for a review, see ref.[2]). In this case, viscous fingering also occurs. Although more complicated than the Hele-Shaw geometry, and more difficult to describe theoretically, the phenomenon is of obvious practical importance in such areas as geology, oil recovery and soil science. As such, it would be useful to characterize the nature of the fingering and its dependence on the physical conditions, e.g. the porosity of the medium, under which it occurs. The present work is the result of a preliminary study to address these questions. Here, we have chosen an approach wherein simulations using Lattice Gas Automata (LGA) are performed and the results subjected to quantitative analysis in order to characterize the resulting patterns and to correlate them with the external parameters.

Notwithstanding its complexity, viscous fingering in a porous medium is ideally suited for LGA simulations for two reasons. First, the problem remains two-dimensional thus allowing us to employ relatively well studied LGA models for the simulations. Second, the nature of the problem, the genesis and evolution of a hydrodynamic instability, requires large simulations which can be run for long times thus taking full advantage of the strengths of the LGA method. In addition, one aspect of the problem we wish to study, the effect of surface tension, requires that we be able to adjust this parameter at will which, as will be described below, is possible within the LGA model that we employ.

Simply performing these simulations provides us with a qualtitative picture of the developement of the viscous fingers[3]. However, in order to understand systematically the role of surface tension, viscosity, porosity, etc. a quantitative characterization is also necessary. This we provide by performing a multifractal analysis of the resulting patterns. In contrast to the Saffman-Taylor experiment, where it is believed that a fractal pattern only occurs in the extreme case of zero surface tension[1], we find the patterns formed in the presence of a porous medium exhibit fractal scaling at low, but nonzero, surface tension. As such, multifractal analysis provides a natural, quantitative characterization of the patterns both as a function of time and as a function of the external parameters.

The present paper is a discussion of work in progress. We will describe the results of two simulations : one involving relatively high surface tension and one in which the surface tension is significantly lower. Both the evolution of the patterns in time and the role of surface tension will be discussed. Specifically, in Section II we describe the LGA model used in our simulations while in Section III we describe the techniques used in our analysis. The results for the two cases mentioned are described in detail in Section IV. We conclude, in Section V, with a brief discussion of our results and our plans for further work.

2 Lattice Gas Model

The cellular automaton model that we have used to simulate the displacement of a highly viscous fluid in a porous medium is based on the two-dimensional two-phase hexagonal lattice gas. In a lattice gas automaton simulation the evolution of a huge amount of fluid particles is tracked, which are moving synchronously with unit speed along the edges of a regular lattice. Particles may collide at the nodes of the lattice, conserving mass and total momentum locally. In [4] it has been shown that in the macroscopic limit such a finite system with discrete dynamics satisfies the continuous Navier-Stokes equations for incompressible athermal flows.

The extension towards two-phase flow modeling involves a coloring of the particles in order to distinguish on a microscopic scale between the phases. This coloring is conserved in the collision. Rothman and Keller have shown how special collision rules can be designed which establish a separation between the two phases with an effective surface tension [5]. The model that we are using is based on the same principle, but it differs in that information about the locality of the phases is propagated across the lattice by the same mechanism as by which the particles are propagated. This simplifies the complexity of the implementation to the level of an ordinary single-phase hexagonal lattice gas, as only strictly local information is needed for the collision.

Also the physical properties of our model are somewhat different. Relatively high values for the surface tension can be obtained over a wide range of average particle densities. Furthermore, individual particles still 'feel' the presence of the interface when they have penetrated the bulk of the opposite color at a distance of a few cells. So, even when the interface is in motion (at a low Mach number) the collision rules can easily maintain the separation. In Fig. (1) a typical configuration of colored fluid particles and 'phase information quanta' on a hexagonal lattice is depicted. A detailed discussion about the design of the collision rules, thereby independently tuning the surface tension and the viscosities of each of the phases, can be found in [6].

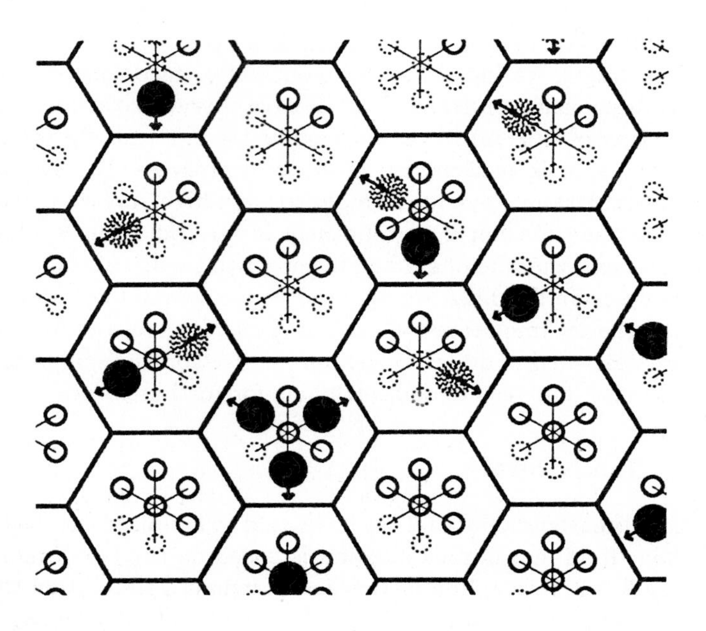

Fig. 1. A typical configuration of the two-phase hexagonal lattice gas. Solid and hatched dots represent the two kinds of particles. Open circles represent positions where no particle is moving (holes). Holes carry a color as well, indicating which phase is most likely found opposite to the direction of propagation.

Basically, there are two approaches by which a cellular automaton can yield the Darcy equations. In both cases the standard evolution of the lattice gas is disturbed by a small fraction (say 1-5%) of scatter points (i.e., nodes at which all particles are bounced backwards during the collision step of the algorithm). Note that this bounce back collision does not conserve momentum, but imposes locally a zero velocity.

Starting from the lattice gas version of the Navier-Stokes equations (in the incompressible limit)

$$\rho \partial_t \mathbf{u} + g(\rho) \rho \mathbf{u} \nabla \mathbf{u} = -\nabla p + \eta \Delta \mathbf{u} \quad , \tag{1}$$

the non-linear term will vanish if the simulation is run at a relative density of

0.5 (such that $g(\rho) = 0$). Now the two alternatives differ in whether the scatter points are allocated to fixed positions in the lattice or whether at each time step an independent random selection is made. If the scatter points are allocated to fixed positions, then the viscous term will dominate any fluctuation in time (in the low Mach number limit). Close to a scatter point the average velocity will be zero and locally relatively steep velocity gradients will persist. On a large scale, the scatter points prohibit different regions of flow to interact otherwise than by means of the pressure. Consequently, the physics is similar to the physics in a Hele-Shaw cell. The solution of (1) is then given by

$$\mathbf{u} = -\frac{\kappa}{\eta}\nabla p \ , \tag{2}$$

where κ is a tensor which depends on the local geometry only.

If scatter positions are selected at random at each time step, the Darcy equations follow directly from the mean field theory of the lattice gas automaton. Consider a fraction α of the collisions in the automaton to be replaced by a (not momentum conservative) bounce back collision. The basic transport equation then shows

$$\begin{aligned} &\partial_t N_i + \mathbf{c}_i \cdot \nabla N_i = \Omega_i \ , \\ &\sum_i \Omega_i = 0 \ , \quad \sum_i \mathbf{c}_i \Omega_i = -2\alpha \sum_i \mathbf{c}_i N_i \ . \end{aligned} \tag{3}$$

In these equations $\mathbf{c}_i$ is the particle velocity, N_i is the average particle distribution and Ω_i is the effect of a collision on the particle distribution. From (3) the Navier-Stokes equations are obtained with an additional term due to the bounce back collisions, i.e.

$$\rho\partial_t \mathbf{u} + g(\rho)\rho\mathbf{u}\nabla\mathbf{u} = -\nabla p + \eta\Delta\mathbf{u} - 2\alpha\rho\mathbf{u} \ . \tag{4}$$

The scatter points, however, are not localized this time and thus no steep velocity gradients will appear on a small scale. Appropriate values for α make the viscous term vanish and the automaton will satisfy

$$\mathbf{u} = -\frac{1}{2\alpha\rho}\nabla p \ . \tag{5}$$

in the incompressible limit. Note, that the viscosity η of the cellular automaton fluid does not appear in this formulation of Darcy's Law! Indeed, only the ratio $\frac{\kappa}{\eta}$ is relevant for the interpretation of the cellular automaton simulation. One should realize, that α may not reach very high values, as this may seriously disturb equilibrium (say $0.001 < \alpha < 0.01$).

We have chosen the first approach to simulate the displacement of a highly viscous fluid in a porous medium. The collision rules have been designed such that the ratio between the viscosities of the two phases is 10.

The pressure in a cellular automaton simulation is in leading order proportional to the density. This complicates the simulation setup quite seriously. If the pressure gradient would be known a priori and would be time-independent (except for small fluctuations), then the pressure field could be decoupled from the density in the simulation by imposing a body force, possibly varying with space. However, in our case the pressure gradient is steep in the viscous parts and almost negligible in the non-viscous parts of the flow and the topology is changing continuously. Therefore, we have to impose the pressure drop explicitly. Furthermore, in order to prohibit non-linearities to develop, we must limit the overall pressure drop and fix the relative density around 0.5. In our simulations we have used a relative density drop of at most 0.2.

A consequence of the above considerations is that in order to increase the size of the lattice by a factor of two in both dimensions (at a constant pressure drop and a constant fraction of scatter points) the pressure gradient and hence the speed of the flow has to be decreased by a factor of two. So, the number of time steps that is needed in order to fully displace the highly viscous fluid will thereby increase with a factor of four and thus the total computing time by a factor of sixteen!

Finite computing resources have enforced us to limit the fraction of scatter points to 1% and the size of the lattice to 2000×800. In this case, the complete evolution of the 1.6×10^6 nodes over 2×10^5 time steps can be completed just within 9 hours on a parallel computer with 400 Transputers. At several instances macroscopic variables, such as local velocity and pressure and the topology of the interface between the phases, have been extracted from the configuration of the particles on the lattice. The following sections present some of the analysis of these data.

3 Multifractal Analysis

Let us assume that we wish to analyze the scaling of some local property, $\rho(\mathbf{r})$, which, for convenience, we take to be normalized so that its integral over the whole system is unity. In our case, $\rho(\mathbf{r})$ will always be the local number density and we will denote the total number of points as N. We begin by considering the so-called partition function, Ξ_q, defined by

$$\Xi_q = \sum_{i=1}^{n(l)} \rho_i^q \tag{6}$$

In eq.(6) we have partitioned the system into $n(l)$ boxes of length l and volume l^d where d is the Euclidian dimension of the system. The integral of $\rho(\mathbf{r})$ over box i is denoted ρ_i. The generalized dimensions, also called the Renyi dimensions are defined by (see, e.g., Ref. 7)

$$(q-1)D_q = \lim_{l \to 0} \frac{\ln \Xi_q}{\ln l} \tag{7}$$

D_0 is the Hausdorff dimension of the object, D_1 the information dimension and D_2 the correlation dimension. However, the variable q clearly need not be integer and D_q may be considered to be a continuous function of q. For large q, the sum is dominated by those boxes with large ρ_i while for q negative it is dominaated by those with the smallest values of ρ_i. If all boxes had the same value of ρ, all of the dimensions would be the same and just equal to the Hausdorff dimension D_0. The generalized dimensions therefore give an indication of how nonuniformly $\rho(r)$ is distributed throughout the system.

A more direct characterization of the distribution of $\rho(\mathbf{r})$ is the spectrum of singularities[7]. Suppose that there is a set of points, $\mathcal{S} = x_i$, such that in the neighborhood of any point in this set $\rho(\mathbf{r})$ behaves as

$$\rho(\mathbf{r}) \sim \mid \mathbf{r} - \mathbf{x}_i \mid^{(\alpha - D_0(\mathcal{S}))} \tag{8}$$

where $D_0(\mathcal{S})$ is the Hausdorff dimension of the set $\mathcal{S}$. We then call α the strength of the singularity in $\rho(\mathbf{r})$. It is clear that $\rho(\mathbf{r})$ could contain many singularities of different strengths so that there might be a whole set of different α's. Let us therefore write $\mathcal{S}(\alpha)$ for the set of all points with singularity α. Then the distribution of singularities throughout the system can be characterized by the (Hausdorff) dimension of the set of points $\mathcal{S}(\alpha)$ which is usually denoted $f(\alpha)$ and called the spectrum of singularities. These quantities, α and $f(\alpha)$, are related to the dimensions via a Legendre transform :

$$\alpha = \frac{d}{dq}[(q-1)D_q] \tag{9}$$

and

$$f(\alpha) = \alpha q(q-1) - D_q \tag{10}$$

For a simple fractal, $\alpha = f(\alpha) = D_0$, the Hausdorff dimension of the object. In the case of a multifractal, the picture is one of interwoven fractals, each of dimension $f(\alpha)$, on which are singularities in $\rho(\mathbf{r})$ of strength α.

The defintions, eqs.(6)-(10) are algorithmic and can be directly implemented to calculate the dimensions and the spectrum of singularities. However, in practice, it is found that this "box-counting" method converges slowly and therefore requires a large amount of data to obtain reliable results. An alternative method, used in the present calculations, is the "correlation integral" approach([8]). Here, the fundamental quantity is the correlation integral, $C(l;q)$ defined as

$$C(l;q) = \frac{1}{N}\sum_{i=1}^{N}[\frac{1}{N}\sum_{j=1}^{N}\Theta(l - \mid \mathbf{x}_i - \mathbf{x}_j \mid)]^q \tag{11}$$

where $\Theta(r)$ is the Heaviside step function and $\mathbf{x}_i$ is the coordinate of point i. The dimensions are then determined as

$$D_q = \lim_{l \to 0} \frac{\ln C(q;l)}{\ln l} \tag{12}$$

The correlation integral therefore counts the fraction of points within a distance, l, of the point $\mathbf{x}_i$ (the inner sum), raises this number to the q-th power and averages this quantity over the whole set (the outer sum). In practice, this procedure is found to converge, with respect to the determination of D_q, much more rapidly than the box-counting method. Heuristically, it is analogous to performing box-counting for many different partitions of the space and averaging the results of all of these determinations of D_q.

4 Application to Viscous Fingering

Using the methods described in Section 2, we have performed LGA simulations of viscous fingering in a porous medium on a universe of 2000*800 nodes for two cases: that of high surface tension, $\sigma \sim 0.280$, and that of moderate surface tension, $\sigma \sim 0.086$. (Simulations in the case of low surface tension will be described elsewhere.) Each simulation was run for nearly 200,000 collision times producing a wealth of data. Our discussion below will be limited to only a few "snapshots" of the systems in order to give an overall view of the evolution in time of the flow. To eliminate some of the noise that an instantaneous snapshot of the systems would include, we have preprocessed the data by averaging temporally over 2500 timesteps and spatially over boxes of 10X10 nodes. Figure (2) shows a typical fingering pattern after this preprocessing.

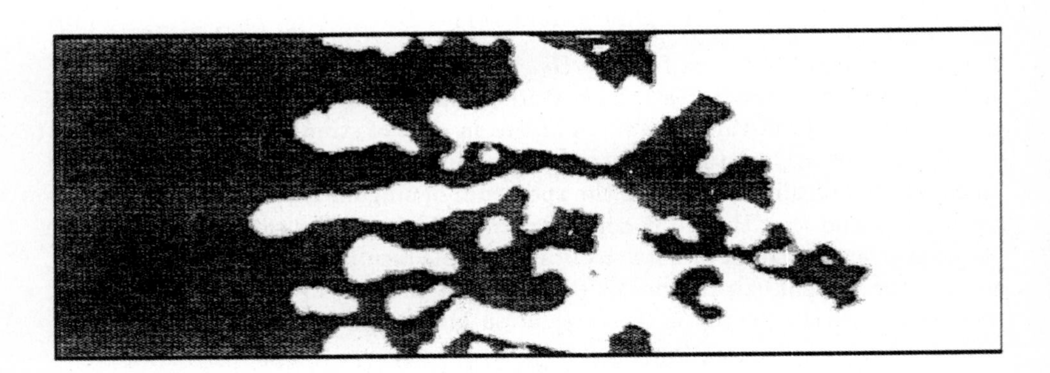

Fig. 2. Viscous fingering in the case of medium surface tension after 100,000 collision times and after temporal and spatial averaging of the data.

In order to characterize the multifractal character of the fingers, care must be taken as to what is actually analyzed. Specifically, in both the high surface tension (HST) and the medium surface tension (MST) simulations, the fingers themselves are two-dimensional in character : rather, it is the *boundaries* of the fingers that we expect to display scaling if it exists at all. The boundary between the fluids is, however, somewhat difficult to identify qualitatively so we have chosen as a simple criterion the following. A pixel (i.e., an average over 10X10 nodes) is considered to be part of the boundary between the fluids if its average concentration is between 0.95 and 0.05 *and* if it has at least one neighbor whose concentration is greater than 0.95. The idea is that since the data is smeared in time, the instantaneously sharp boundary between the fluids is also smeared. We therefore pick out an edge of the boundary by requiring that there be "pure" neighbors. Comparison between plots of our data and plots of the boundary based on this criterion indicate to us that the criterion is reasonable.

Having identified the boundary, we proceed to attempt to determine the generalized exponents using the correlation integral method. To establish whether or not scaling exists at all, we first plot the correlation integral, $C(q;l)$, vs. l; if scaling exists, the plot should be linear. In Figs.(3) and (4) we show plots for both HST and MST for q=-3,0 and 3 after 100,000 time steps. We are restricted to length scales between 1 and 30 pixels, the upper limit coming from the fact that the system is periodic and we must avoid counting a point as its own neighbor. In fact, no physical object is ever "truely" fractal since at very small length scales, both stochastic noise and the underlying structure of the object (atomic composition, the presence of a lattice, etc.) destroy scaling. Thus, one searches for a range of length scales over which the corellation integrals depend linearally on l. (In the present case, we never go below $l = 5$ pixels.) Surprisingly, the HST correlation integrals appear to show scaling corresponding to the rather large Hausdorff dimension of 1.41. However, on closer inspection, a systematic trend in the data is apparent and if we perform a quadratic fit, we find that the slope at $l = 0$ extrapolates to 1.05 which indicates no (significant) anomolous scaling in this case. It is interesting, however, that at larger length scales, the slope is so large. This is not surprising since one expects that the strong surface tension forbids "tip-splitting" at small length scales where it would give rise to a large curvature of the interface whereas at large length scales tip-splitting can, and does, occur thus giving rise to self-similarity or scaling at these length scales. At later times, the lower limit, in l, of the scaling decreases leading to the possibility that the pattern is slowly becoming fractal. Unfortunately, we only see scaling (i.e., a linear relation between $C(q;l)$ and l) after about 150,000 time steps when we have only a few hundred boundary points because most of the high viscosity fluid has been pushed out of the simulation cell. This makes a reliable determination of the scaling properties at this time very difficult. It is worth noting, however, that we calculate a Hausdorff dimension at t=150,000 of 1.46 which is consistent with that at t=100,000 discused above.

As can be seen in Fig.(4), we appear to have good scaling in the MST simulation. Comparing linear and quadratic fits give the same slope, to within 3%, thus indicating that the scaling is real. We have therefore been able to calculate both the dimensions and the spectrum of singularities in this case. Figure (5) shows the generalized dimensions at t=50,000, 100,000 and 150,000 collision

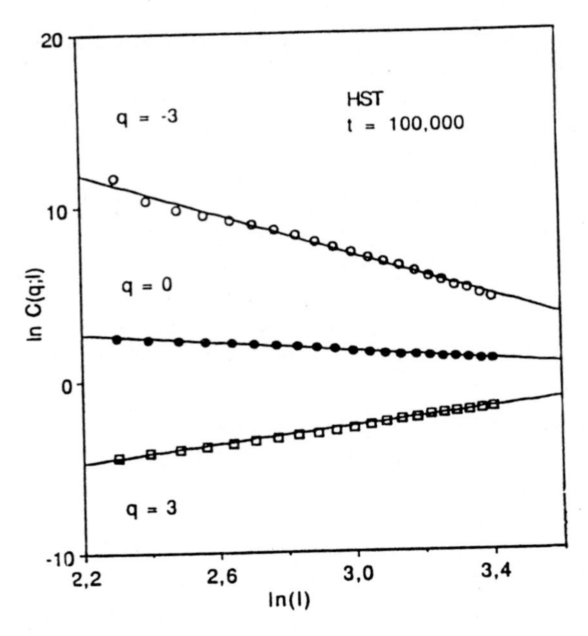

Fig. 3. The logarithm of the correlation integrals, C(q;l), as a function of $\ln l$ for q = -3, 0, and 3 for HST.

times. The error bars shown are based on an estimated error of twice the standard deviation of the coefficients of a linear fit to the data. At negative values of q, the error increases since, as previously mentioned, the correlation integrals are dominated by those points with fewest neighbors which are most subject to statistical fluctuations. At early times, $D_q = 1$ for all values of q because the initial interface is planar. However, already at t=50,000 we see that considerable structure has developed as is indicated by a Hausdorff dimension of 1.61 and the width of the distribution of dimensions. At 100,000 collision times, the system is already nearing a steady state. The Hausdorff dimension of about 1.75 does not change over the next 50,000 time steps and the dimensions at negative q are stationary to within the estimated error in their determination. Only in the positive q branch do we see a significant change. Since these values are dominated by the points with the most neighbors, and since D_q increases with time, we conclude that this is due to a relatively slow densification which is probably due to further tip-splitting at small length scales.

The spectra of singularities are shown in Fig.(6) where the same trends are apparent. At t=50,000, we find a rather broad spectrum indicating a wide range of growth rates within the system whereas at later times the spectrum narrows as the system stabilizes. These spectra, particularly at the latest times, are quite similar to those observed in simulations of diffusion limited aggregation (DLA) and in viscous fingering experiments with nonNewtonian fluids having negligable surface tension[9]. (Indeed, it is conjectured that viscous fingering (without a

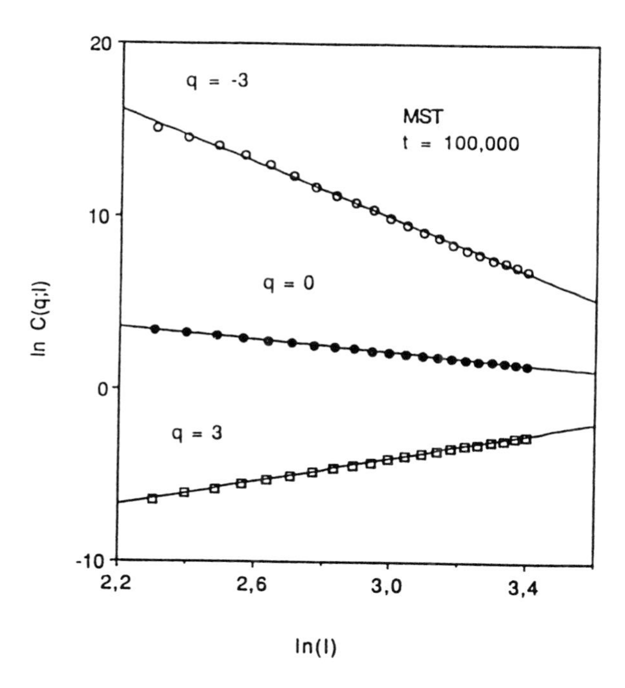

Fig. 4. The logarithm of the correlation integrals, C(q;l), as a function of $\ln l$ for q = -3, 0, and 3 for MST.

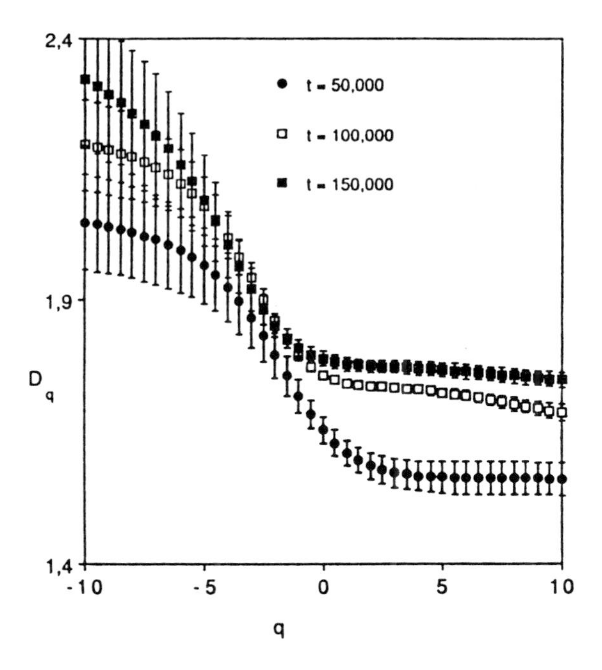

Fig. 5. The generalized dimensions, D_q, for the case of MST.

porous medium) is isomorphic to DLA in the limit of zero surface tension[1].) In particular, our Hausdorff dimension of 1.75 is remarkably close to that seen in DLA (about 1.7). The spectra differ in that in our case, the range of values of α for which the spectrum is defined is a rather modest 0.8 whereas in DLA the range is typically about 10 times this while experiment gives about 5 times our range. This can be understood as follows. In the absence of a porous medium, we would not expect to see any anomolous scaling at finite surface tension. The presence of the medium therefore has the effect of destabilizing the interface and giving rise to the observed scaling. Even so, only a relatively narrow range of singularities is observed in our MST case because of the surface tension. It might be conjectured that the Hausdorff dimension is relatively insensitive to the physical parameters but that the range of singularities is quite sensitive to them.

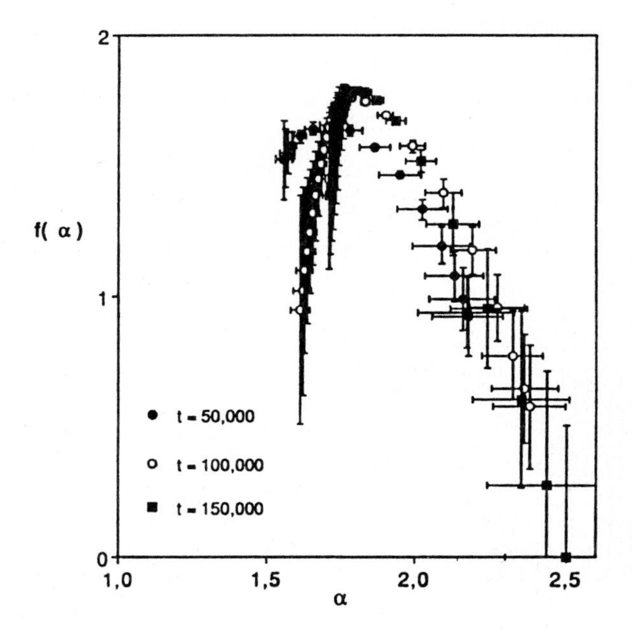

Fig. 6. The spectra of singularities calculated for the case of MST from the dimensions shown in Fig.(5).

5 Discussion

The results of a multifractal analysis of LGA simulations of viscous fingering in a porous medium have been described. We have found that at high surface tension, the patterns are not definitively fractal in nature although it is possible that they become so very slowly. With a moderate surface tension, the patterns are clearly multifractal with a Hausdorff dimension similar to that found in DLA. The spectrum of singularities is however much more narrow than in either DLA or Hele-Shaw type experiments indicating a more uniform growth pattern than in these cases. We believe that as the surface tension is lowered, the spectrum should become wider. In addition, it is reasonable to imagine that a larger porosity would broaden the spectrum since the porous medium is the destabilizing factor. We are presently investigating these conjectures as well as the influence of the geometry on the growth.

Acknowledgements

JPB acknowledges support from the Fonds National de la Recherche Scientifique (FNRS, Belgium). This work was supported by EC grant SC1-0212.

References

1. David Bensimon, Leo P. Kadanoff, Soudan Liang, Boris I. Shraiman, and Chao Tang, " Viscous Flow in Two Dimensions", *Rev. Mod. Phys.* **58** (1986) 977.
2. G. M. Homsey, " Viscous Fingering in Porous Media", *Ann. Rev. Fluid Mech.* **19** (1987) 271.
3. Preliminary LGA simulations are presented in M. Bonetti, A. Noullez and J. P. Boon, "Lattice Gas Simulation of 2-D Viscous Fingering", in : Cellular Automata and Modelling of Complex Physical Systems, edited by P. Manneville, et al. (Springer-Verlag, Berlin, 1989).
4. U. Frisch, B. Hasslacher and Y. Pomeau,"Lattice Gas Automata", *Phys. Rev. Lett.* **56** (1986) 1505-1508.
5. D.H. Rothman and J.M. Keller, "Immiscible Cellular-Automaton Fluids", *J. Stat Phys.* **52** (1988) 1119.
6. J.A. Somers and P.C. Rem, "Analysis of Surface Tension in Two-phase Lattice Gases" *to appear in Physica D* (1990)
7. Thomas M. Halsey, Mogens M. Jensen,Leo P. Kadanoff, Itmar Procaccia and Boris I. Shraiman, " Fractal Measures
8. See, e.g., T. Vicsek, " Mass Multifractals", *Physica A* **168** (1990) 490.
9. H. Eugene Stanley, and Paul Meakin, " Multifractal Phenomena in Physics and Chemistry", *Nature* **335** (1988) 405.

This article was processed using the LaTeX macro package with ICM style

DARCY FLOW IN POROUS MEDIA: CELLULAR AUTOMATA SIMULATIONS

J.M.V.A. Koelman and M. Nepveu
Koninklijke/Shell Exploratie en Produktie Laboratorium
P.O. Box 60, 2280 AB Rijswijk, The Netherlands

ABSTRACT

This contribution presents two cellular automata for the computation of 2-D, incompressible, one- and two-phase Darcy flow. The one-phase automaton is based on a Navier-Stokes automaton (Frisch, et al.). Darcy characteristics are brought in with the help of so-called scatter cells. The convergence is speeded up by a kind of multi-grid technique. The two-phase automaton is based on diffusion-limited aggregation (DLA). In this automaton 'voids' are released at a production well, they make their way to an injection well and then leave the system. The motion of the voids, which is controlled by the mobility of the fluids and by a stochastic parameter, stirs the fluids and determines the flow.

The results obtained so far suggest that work on more realistic situations is particularly promising along the lines of the DLA automaton.

1. INTRODUCTION

Knowledge of the way oil, gas and water flow in the reservoirs from which oil or gas are produced is of paramount importance in the oil industry. Because a lot of research in this connection is focussed on how to calculate Darcy flow of these components at a reservoir scale, it is apt that some of this research be directed to the application of the relatively new technique of cellular automata. It is reasonable, then, to begin with an inquiry into cellular automata for one-phase flow so as to learn about the pros and cons of the method. The ultimate goal, however, must be the construction of a Darcy cellular automaton (CA) capable of handling compressible, 3-D multi-phase flow in a gravitational field. Hence, there is still a long way to go after these initial steps!

In this contribution we start with a CA for one-phase Darcy flow. The CA is based on the Navier-Stokes automata of Frisch, et al. (Ref. 1) that have been adapted so as to produce Darcy flow by means of scatter cells in the grid. These scatter cells mimic in some sense the many obstacles (interstices and walls) the fluid will encounter in the porous medium in which it flows. This CA 'narrows down' the Navier-Stokes solutions to those pertaining to reservoir conditions.

On the other hand, one can forget about Navier-Stokes automata altogether and start with rules that seem promising for computing two-phase flow. This method is still heuristic; eventually a proof must be furnished of the viability of these new rules.

The organisation of the paper is as follows: in Section 2 our one-phase Darcy CA is presented; in Section 3 a two-phase automaton is described and in Section 4 these CA's are discussed and their interrelations are elucidated. Section 5 offers some conclusions.

2. ONE-PHASE DARCY CA

We define a hexagonal lattice and introduce particles in the lattice cells. As many as seven particles can coexist in any one cell - one for each direction of motion plus a 'rest particle'. At each step in the 'grinding' of the CA, the particles will move to adjacent cells (except the rest particle) and thus redistribute themselves over the lattice. After this translational motion over the distance of one cell, the particles interact with each other by means of collisions so that the total number of particles and the total momentum per cell remain constant. We apply the so-called FHP-II rules defined in Ref. 1 to do this. Then the state at time j+1 in a cell S(j+1) is related to the cell's previous state S(j) by the symbolic formula S(j+1) = (CT) S(j), where T is the translation operator (which depends on neighbouring cells) and C is the collision operator. So each new state of a cell follows from the one at a previous time by local rules. If one now averages the particle numbers and the momenta over (sufficiently large) connected areas, these averages give the **pressures** and **velocities** of **2-D incompressible Navier-Stokes flow** under conditions given explicitly in Ref. 1.

We model Darcy flow by introducing 'scatter cells' into the lattice. These cells bounce a particle back into the very cell where it came from. In Ref. 2 it was proved (for a special case) that Darcy flow ensues, if the density of the scatter cells is sufficiently high that they make themselves felt. (A precise criterion of how high that is has not yet been formulated.) On the other hand, if the scatter cells are sufficiently dispersed so that no velocity correlations are built up between them, a simple relation can be derived between the scatter-cell density <W>, the viscosity υ, and the permeability k. When the system has relaxed to a steady state (which is presumed to exist for the system under scrutiny) this relation is: $k = \upsilon/2\langle W\rangle$. Hence, by varying the density of the scatter cells over the lattice, high- and low-permeability regions can be mimicked.

In order to reach a steady state, one has to run this CA for a number of sound travel times (the number of lattice cells in the length direction divided by latticed sound speed). For runs on a grid of 300x600 cells, as we used, this would require waiting for, say, 3000 time steps. We can do better, however, if we implement a neat little trick, which is essentially a multi-grid technique (Ref. 3). For incompressible one-phase Darcy flow the pressure field in a steady state is governed by the (generalised Laplace) equation div (k/υ)grad p = 0. This equation is solved with relevant boundary conditions on a coarse grid, for instance with a finite-difference method. The solution on the coarser grid can then be transferred to the much larger CA grid as the initial condition for the pressure field. (We do not provide an initial guess for the velocities; the CA must accommodate these all by itself.) Thus, the final steady-state solution for p is already approximated even before the CA is run. We surmised that the CA would achieve a steady state much quicker with this friendly help.

Figure 1 shows a 12x24 grid on which the generalised Laplace equation is solved. The permeability in the top-right and bottom-left corners is 11 times lower than elsewhere on the grid. Pressures are fixed at the inlet and outlet. No flow is allowed over the remaining boundaries (**n**.grad p = 0, where **n** is the boundary normal). In Figure 2a the pressure field is given after 50 time steps. In Figures 2b-e the velocity field is plotted after 50, 150, 250 and 350 time steps. Clearly, convergence is reached well within the lattice sound travel time. This speed-up saves computer costs. Moreover, it helps to establish meaningful solutions quickly when boundary conditions are changed abruptly in a problem. This trick may also be very helpful in higher-dimensional computations. It offers advantages in starting-up multi-phase flows (in more advanced CA's) as well, although steady state will not normally exist in such cases.

3. TWO-PHASE DARCY CA

We have also devised a novel automaton for the simulation of two-phase miscible Darcy flow. Although at first sight this scheme seems to be the result of a complete departure from the above FHP-based automaton, it may be viewed as the result of stripping the "Navier-Stokes + scatter-cell automaton" to its barest Darcy essentials. Alternatively, the scheme may be viewed as a generalisation of the DLA (diffusion-limited aggregation) algorithm originally devised by Witten and Sander to model particle-cluster aggregation (Ref. 4). Both relationships will be elucidated in the next section. For the time being, we restrict ourselves to the scheme itself and the results obtained with it.

The automaton has no more than three states per cell. These states relate to whether or not fluid is present in a given cell, and, if so, which of two fluids occupies the cell. In view of the problems we want to address, we refer to cells in the various states as: "filled with reservoir fluid", "filled with injection fluid" and "empty". The dynamics originates from the fact that at every "tick of the clock" both fluids try to move to a neighbouring cell. The fluids do so at a rate proportional to their mobilities. In order to prevent a cell from becoming filled with more than one fluid, these trial moves are disregarded when the target cell is already filled with fluid. As a result, only those filled cells that are adjacent to an empty cell participate in the dynamics, which essentially involves replacing the empty cell with one of the adjacent filled cells. The replacement is decided randomly, with probabilities proportional to the different mobilities of the fluids occupying the different cells. The system is driven by injecting "voids" into the cells where fluids can leave (production sites). The number of voids in the system stabilises because the voids disappear (become filled with injection fluid) at the cells where fluid is injected (injection sites).

The first simulations have focussed on 2-D systems. As may be clear from above, however, the generalisation to 3-D systems is straightforward. We used a square lattice in our simulations. (Notice that the tensors of fourth order occurring in the Navier-Stokes equations, which necessitate the use of a hexagonal lattice, do not occur in Darcy's equations.) Figure 3 shows the type of patterns that are produced in such simulations when the injected fluid has a much higher mobility then the reservoir fluid. The bypassing of large regions of reservoir fluids through the formation of ramified patterns of injected fluid, commonly referred to as viscous fingering, is one of the causes of reduced recoveries during waterfloodings of oil reservoirs.

More quantitative results are obtained by averaging the sweep efficiencies obtained in a number of independent simulations on a quarter of a so-called five-spot well pattern (Figure 4). This geometry is chosen because the five-spot problem is a classical test for reservoir simulators for which experimental results are available in the literature. We first examined the dependency of sweep efficiencies on grid dimensions for a fixed mobility ratio. The results are shown in Figure 5. This figure gives the fraction of reservoir fluid remaining after injecting a volume of more mobile fluid equal to 0.5 and 1.0 times the total pore volume. As can be seen when expressed in terms of fractional recoveries, the simulation results for the sweep efficiencies of the displacement processes are independent of lattice size (except for very small lattices).

Having established that our CA results can be made to be independent of lattice size, we were able to predict fractional recoveries as a function of mobility ratio and volume of injected fluid and to compare them with results from laboratory experiments. The three curves in Figure 6 show the simulation results for mobility ratios (mobility of invading fluid divided by mobility of injected fluid) equal; to 71.5, 38.2 and 17.3. These values are chosen so as to coincide with the three highest mobility ratios for which Haberman gives experimental results (Ref. 5). As is clearly visible, the simulation results correlate very well with the experimental results.

4. RELATIONS BETWEEN VARIOUS MODELS

As already mentioned, the CA models discussed above are related to each other, and the two-phase CA is closely related to the DLA growth model. In order to see the first connection, we "strip" the one-phase CA to its barest essentials. First, we observe from the theory in Section 2 that the only feature that counts in obtaining Darcy behaviour is the randomisation of momentum in one-particle collisions at the scatter cells. So, all the processing taking place at non-scatter cells is in a certain sense superfluous. A much more efficient algorithm would be possible if we were able to increase the scatter-cell density, without introducing correlated re-collisions. To be able to do so, however, it is necessary to introduce some randomness in the directions in which particles entering a scatter cell scatter. In fact, if the directions are simply chosen at random, we could simplify the automaton by increasing the scatter cell density to unity. Having only scatter cells, the momentum-conserving collisions between particles at the same site are superfluous, and therefore the scheme can be simplified by allowing at most one particle per cell. Furthermore, if momentum conservation is completely dropped, the fourth-order tensor in the momentum-conservation equation disappears, and the hexagonal lattice might as well be replaced by a square lattice. In this way we arrive at a stochastic automaton that is the one-phase version of the CA described in Section 3.

The connection of the two-phase flow automaton with the DLA algorithm (Ref. 4) becomes evident by following the dynamics of the voids. It is easily seen that the stochastic dynamics described in Section 3 leads to voids performing a random walk: the voids move stochastically to neighbouring cells with a probability proportional to the mobility of the fluid occupying those cells. Such a "void jump" to a neighbouring cell is tantamount to a flow of the fluid initially occupying the neighbouring cell to the cell that was originally a void. In the limiting case where the injection fluid is infinitely more mobile then the reservoir fluid and the void density is small (the latter always being the case in our simulations), this dynamics reduces to the DLA process. The reason is that as soon as a void moves to a cell neighbouring the aggregate of injection fluid, it is bound to enter the aggregate in the next step and never leave it until it ultimately reaches an injection cell, where it is "eaten away" by the injected fluid.

5. CONCLUSIONS AND OUTLOOK

In this contribution we have presented a one-phase Darcy automaton based on a Navier-Stokes automaton. We also presented a two-phase Darcy automaton that was not based on Navier-Stokes but employed local rules which are much simpler, though still physically justified. The link between the two is the idea that 'obstacles' must be introduced into the flow field in order to obtain Darcy flow. In the Navier-Stokes-based automaton special (sparse) scatter cells perform this task. In the DLA automaton **each** cell is a scatterer and the scattering can happen in **all** directions. This latter approach enables the handling of two phases 'for free', whereas in the first automaton inclusion of two phases would require large adaptations. Another indication of the flexibility inherent in the DLA automaton is its easy extension to three dimensions, whereas such an extension again requires substantial adaptations to Navier-Stokes-based automata.

The DLA automaton, then, looks like a promising subject for further research. In particular, one should look into the possibility of providing pressure information to the automaton in such a way that it can be incorporated into the CA rules. We are currently investigating this approach. As indicated in the introduction, more physical effects should also be introduced (gravity and relative permeabilities, for instance). The flexibility of the DLA automaton may offer enough scope for this research to be successful.

REFERENCES

1) U. Frisch, D. d'Humieres, B. Hasslacher, P. Lallemand, Y. Pomeau, J.-P. Rivet, Complex Systems **1**, 649 (1987).
2) K. Balasubramanian, F. Hayot, W.F. Saam, Phys. Rev. A **36**, 2248 (1987).
3) W. Hackbush, U. Trottenberg, Multigrid Methods, Springer Verlag, Berlin 1982.
4) T.A. Witten and L.M. Sander, Phys. Rev. Lett. **47**, 1400 (1981).
5) B. Haberman, Trans. AIME **219**, 264 (1960).

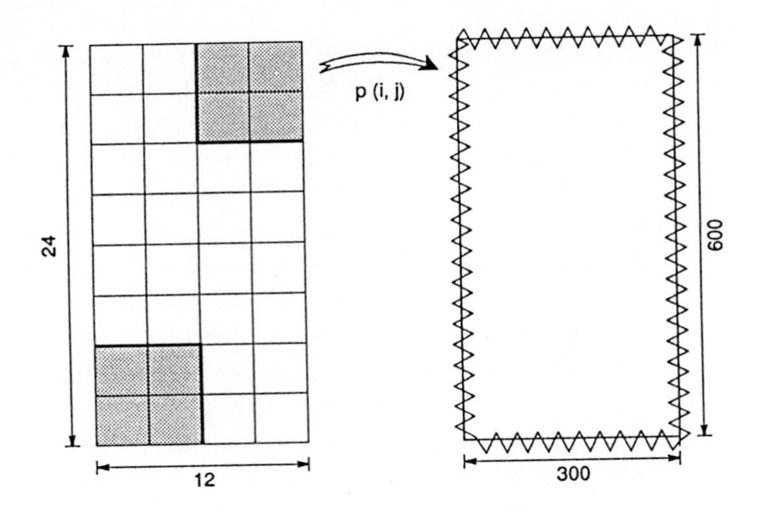

Fig. 1 Pressure information on a 12 x 24 grid is carried over onto the CA grid of 300 x 600. Shaded areas have a permeability 11 times lower than elsewhere

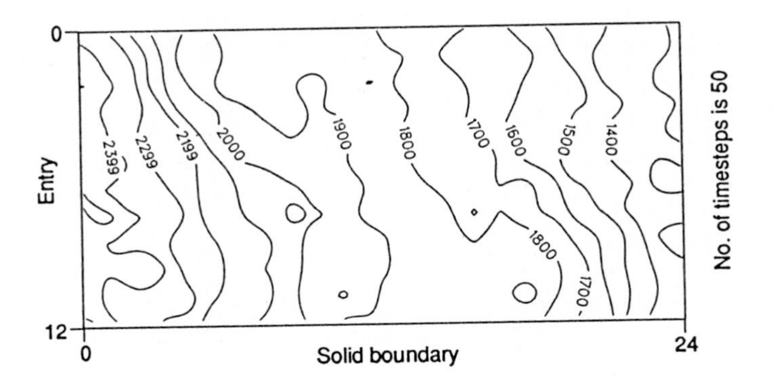

Fig. 2a Pressure contours after 50 time steps

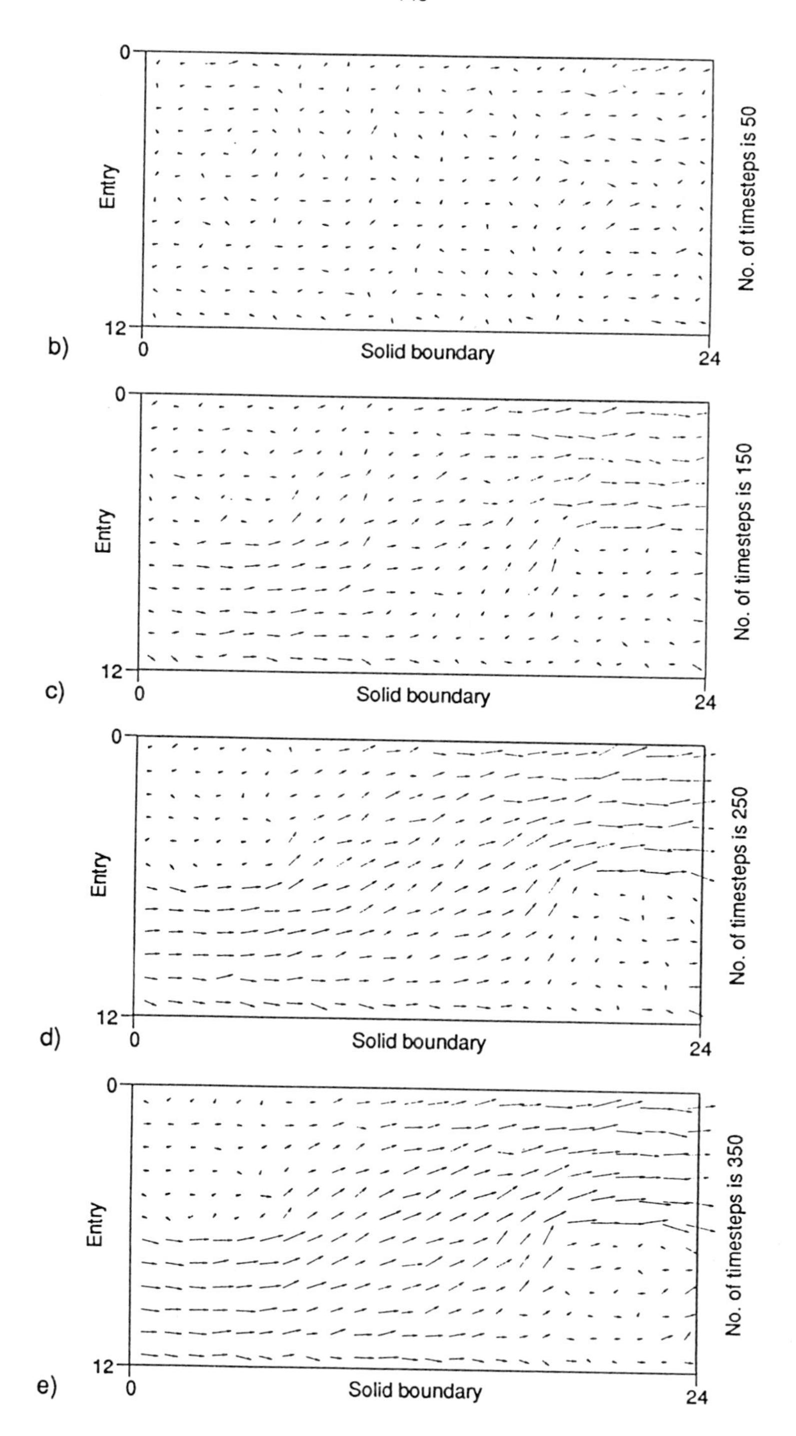

Figs. 2b-e Flow field "in the making". Convergence is reached after about 250 time steps. Sound travel time over the lattice: 1000 steps.

Fig. 3 Simulation of viscous fingering: a high-mobility fluid (entering from below) displacing a much less mobile fluid (leaving from above). The simulation is done on a 256 x 256 lattice.

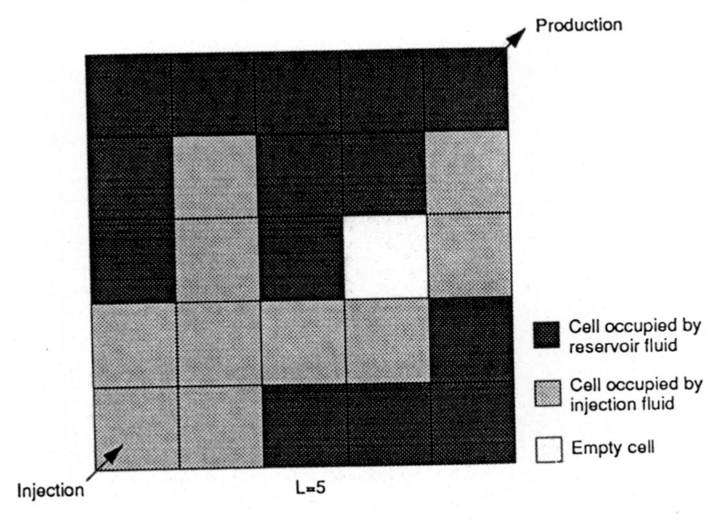

Fig. 4 Quarter of a five-spot well pattern (realisation on a 5 x 5 lattice)

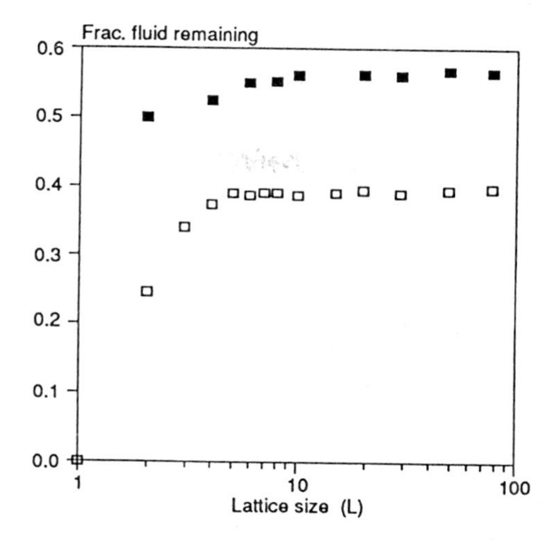

Fig. 5 Fraction of reservoir fluid remaining in a quarter of a five-spot well pattern, as a function of linear lattice size, after injecting a given volume of fluid with 30 times greater mobilty than reservoir fluid. Filled squares: half a pore volume injected; open squares: a full pore volume injected.

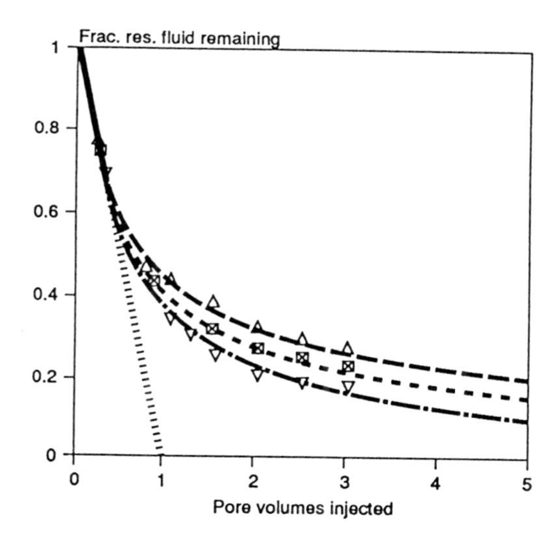

Fig. 6 Fraction of reservoir fluid remaining in a quarter of a five-spot well pattern as a function of the volume of fluid injected (expressed in units of total pore volume). The three curves give the simulation results obtained on a 40 x 40 lattice. Upper curve: mobility ratio M = 71.5; middle curve: M = 38.2; lower curve: M = 17.3. The symbols indicate experimental values obtained from the experiments by Habermann for the same mobility ratios. For comparison, the line corresponding to a displacement with 100 % efficiency is also indicated (light dotted line).

CELLULAR-AUTOMATA-BASED COMPUTER SIMULATIONS OF POLYMER FLUIDS

J.M.V.A. Koelman
Koninklijke/Shell Exploratie en Produktie Laboratorium
Rijswijk, The Netherlands (Shell Research B.V.)

ABSTRACT

A lattice-gas model is constructed that incorporates excluded-volume interactions between the particles and allows for flexible multiparticle structures. With this model, a flexible polymer in a sea of solvent is simulated. As a result, static and dynamic scaling behaviour is observed for the first time in a simulation that takes both excluded-volume and hydrodynamic interactions into account.

INTRODUCTION

One of the main applications of polymers is in controlling the transport properties of fluids: very small amounts of polymers added to a fluid can change the viscosity and other transport coefficients dramatically. They are therefore used in many industrial applications including various oil production operations.

However, a computational tool for predicting the flow properties of polymer solutions from first principles is still lacking. In predicting such properties one has to face the formidable problem of solving hydrodynamics in the presence of random flexible structures. A straightforward molecular dynamics (MD) attack on this problem is beyond the limit of what is computationally possible at present and in the foreseeable future. The main reason for this is that to observe asymptotic scaling behaviour one has to simulate large polymers with many degrees of freedom in a sea of solvent with an even greater number of degrees of freedom. Moreover, because configurational relaxation times increase sharply with polymer length, the system must be monitored over a large time span to sample a sufficient number of uncorrelated configurations. The combined result is that one has to follow these large systems over periods of time that are, compared with MD standards, truly gigantic.

Cellular-automata-based simulations might provide a promising new route for solving this problem. However, the standard lattice-gas automata capable of simulating Navier-Stokes hydrodynamics[1] lack two features that must be included to enable structurally complex fluids to be simulated. Firstly, macromolecular structures must be incorporated in the scheme, i.e., aggregates of particles with a structure-preserving dynamics. Secondly, and perhaps more fundamentally, there is the problem that while some people tend to look upon the standard lattice-gas automata as a kind of "poor man's molecular dynamics", one of their drawbacks is that particle trajectories cannot be followed because the particles have no size and lose their identity in collisions. Although particle identity might be restored (for instance, stochastically) by relabelling the particles in post-collision states[2], this does not lead to a completely satisfactory model because excluded-volume effects, which profoundly affect the dynamics of macromolecules in fluids, are still missing. What one really needs is the notion of hard-core interactions[3] in lattice-gas models preventing the particles from occupying the same site.

In the next section a way in which both hard-core interactions and flexible macromolecular structures can be incorporated in a lattice-gas model are discussed. The first results obtained with this model are then presented, followed by a discussion on ideas for further research. A preliminary report on the material presented in this contribution is given in Ref. 4.

THE MODEL

To incorporate excluded-volume interactions and flexible structures in a lattice-gas model in a simple and consistent way, a number of lattice updates are made at each time step ("tick of the clock"). A time step starts with a collision iteration. For each iteration step (Fig. 1a-b, b-c, c-d) and at each lattice site the number of particles having that site as target site (i.e. that would arrive at that site by free propagation in a unit time step) are counted. If a certain site is the target of two or more particles, these particles collide, i.e., they exchange momentum according to a definite local rule. No action is taken when a site is the target of a single particle or of none of the particles. Such a collision scan over all target sites is repeated over and over again. This iteration ends when none of the target grid sites of any two particles coincide. The computer simulations discussed below show that, for particle densities of up to about 50% site occupancy, this iterative scheme of collisions always converges rapidly to the desired final state in which all particles have their velocity vectors pointing to different sites. For higher densities, convergence is still reached, but only after a larger number of iterations. Note that, because in this iterative process particles having different target sites do not interfere with one another, the order in which the target sites are processed in a lattice update does not matter, and in fact can be done in parallel. Also note that in this collision iteration the target grid sites of all the particles are continuously updated, but the particles themselves do not move.

The collision iteration is followed by a chain-velocity-update phase. In this phase the velocities of the particles in a chain are updated to prevent particles that are neighbours in the polymer chain from moving too far apart. This is done by redistributing the velocities of the particles in a chain (Fig. 1d-e) using trial velocity exchanges to minimise the sum of the squares of the distances between the target sites of neighbouring chain particles, without violating the condition that no target grid sites may coincide. The chain-update phase ends when no further velocity exchange attempts are successful.

The last phase is simply a free propagation of all the particles to their actual target grid sites (Fig. 1e-f). In the resulting configuration the excluded-volume condition (no multiple occupancy of any grid site) is strictly met, and the chain structures are preserved. Given that the number of particles, linear momentum and kinetic energy are collision invariants (i.e., conserved during the collision iteration), these are automatically conserved throughout the whole time step (all three phases).

These 'rules of the game' are quite general and do not rely on the use of a particular regular lattice of sites, a specific set of velocity vectors or a definite choice of collision rules. For the 2-D studies a square lattice is chosen with a corresponding set of nine velocity vectors: one zero velocity (rest particles), four unit velocities (particles moving to nearest neighbour sites), and four $\sqrt{2}$ velocities (particles moving to next-nearest neighbour sites). This choice has the definite advantage that it allows for collision rules that conserve the number of particles, momentum and kinetic energy independently of one another. Choosing such collision rules yields a model having a certain thermal property[5]: a non-trivial conservation of kinetic energy.

The simulation results to be discussed below show that this specific choice of lattice and set of velocity vectors does not yield any anisotropy problems for the quantities studied. For studies in 3-D a straightforward extension of this model is used: a cubic lattice with a set of 19 velocity vectors[5] (one zero velocity, six unit velocities, and twelve $\sqrt{2}$ velocities) and collision rules that independently conserve the number of particles, momentum and kinetic energy.

RESULTS

The first simulations have focussed on polymers in a 2-D sea of solvent (Fig. 2). Quantitative information is obtained for the radius of gyration (R_g) and the centre-of-mass diffusion constant (D) of polymer coils when only one polymer is present (infinite dilution). A clear scaling behaviour – a power-law dependence on the degree of polymerisation – is observed for both quantities. The scaling behaviour shown in Fig. 3

agrees very well with the theoretical prediction $R_g \sim N^{0.75}$ for self-avoiding random walks in 2-D. Of course, hydrodynamic interactions do not affect this static scaling behaviour. Of much more interest in this respect is the scaling behaviour observed for the centre-of-mass diffusion constant (Fig. 4). This dynamic scaling is very profoundly influenced by hydrodynamic interactions, and it is in this quantity that the strength of the cellular-automata (CA) scheme shows up. As can be clearly seen in Fig. 4, the CA model predicts a dynamic scaling $D \sim N^{-0.78\pm0.05}$. To my knowledge, this is the first time that such a dynamic exponent has been determined by a computer simulation that takes into account the fluctuating hydrodynamic interactions. The observed dynamic scaling behaviour contrasts sharply with the scaling behaviour expected from the commonly accepted[6] non-draining picture. In this picture hydrodynamic interactions are assumed to dominate in such a way that the polymer coil drags along the solvent inside it. As a result the polymer coil and the fluid inside it act together as an impermeable object around which the solvent has to flow. Clearly, in this limit the dynamic behaviour of the polymer should scale as that of a sphere with radius equal to the radius of gyration of the polymer. In 2-D, non-draining implies that the diffusion coefficient of the polymer should scale like that of a disk and hence should depend only logarithmically on polymer length. The observed scaling behaviour clearly violates the non-draining assumption.

This anomalous scaling behaviour is caused by hydrodynamic effects and not by some artifact in the model. This can be inferred from simulations with the same model in which the hydrodynamic interactions are "turned off". This is done by extending the set of allowed collisions by permitting collisions that still conserve the number of particles and the kinetic energy, but which violate momentum conservation. The results of these simulations are also shown in Fig. 4 (open symbols). Clearly these simulations lead to a scaling behaviour of the polymer diffusion constant described by $D \sim 1/N$. This result corresponds to the free-draining or Rouse limit which results from theories that ignore hydrodynamic interactions[6].

The fact that experimental 3-D results also show deviations from non-draining behaviour[7,8] suggests that the anomalous dynamic scaling behaviour observed here is not limited to 2-D systems. Deviations in 3-D experiments are observed only when excluded-volume effects are important[8] (good solvents), this suggests that such deviations probably result from a subtle interplay between excluded-volume and hydrodynamic effects. This also offers a possible explanation for the fact that the deviations from non-draining behaviour observed in our 2-D simulations are much larger than the corresponding experimental 3-D deviations[7,8]: by reducing the number of spatial dimensions to two, both excluded-volume and hydrodynamic interaction effects are increased.

Full 3-D simulations should shed more light on the problem of the hydrodynamic behaviour of polymers that are "swollen" as a result of excluded-volume effects. Here, the first 3-D results obtained with the lattice-gas model are reported. As a first exercise in 3-D we focus on the static scaling that can be observed in the dependence of the radius of gyration on polymer length (Fig. 5). As is clearly visible in this figure, the theoretical scaling relation $R_g \sim N^{0.59}$ is confirmed. The investigation of dynamic scaling requires a larger amount of computational effort, and is left for further research.

OUTLOOK

It has been shown in this contribution that lattice-gas automata can be extended to model polymer fluids. Although the quantitative results discussed here are limited to a single flexible polymer in a sea of solvent, the method is much more general. There is no problem in modelling more than one polymer in a solution (Fig. 2), so one is not restricted to the infinite dilution limit. Furthermore, besides linear polymer chains, ring polymers, star polymers, polymer chains with side branches, polymer networks and any mixture of them can be modelled. In fact, flexible macromolecular structures with arbitrarily complex topology can be incorporated. Of course, one is not limited to the measurement of diffusion constants, other transport coefficients can also be determined. Simulations under non-equilibrium conditions are of particular interest; for instance, the effect of imposing certain flow fields (shear flow, elongational flow, etc.) on the rheological behaviour of various different polymer models can be compared.

These points are all topics for further research. Of crucial importance for future research is, of course, the construction of efficient 3-D models that do not require excessive amounts of computational effort for realistic simulations.

REFERENCES

1. U. Frisch, B. Hasslacher,, and Y. Pomeau, Phys. Rev. Lett. **56**, 1505 (1986).
2. D. Frenkel and M.H. Ernst, Phys. Rev. Lett. **63**, 2165 (1989).
3. M.E. Colvin, A.J.C. Ladd, and B.J. Alder, Phys. Rev. Lett. **61**, 381 (1988).
4. J.M.V.A. Koelman, Phys. Rev. Lett. **64**, 1915 (1990).
5. D. d'Humieres, P. Lallemand, and U. Frisch, Europhys. Lett. **2**, 291 (1986).
6. P.G. de Gennes, Scaling Concepts in Polymer Physics (Cornell Univ. Press, Cornell, 1979).
7. M. Adam and M. Delsanti, J. Phys. (Paris) **37**, 1045 (1976); M. Adam and M. Delsanti, Macromolecules **10**, 1229 (1977).
8. A.R. Ellis, J.K. Schaller, M.L. McKiernan, and J.C. Selser, J. Chem. Phys. **92**, 5731 (1990).

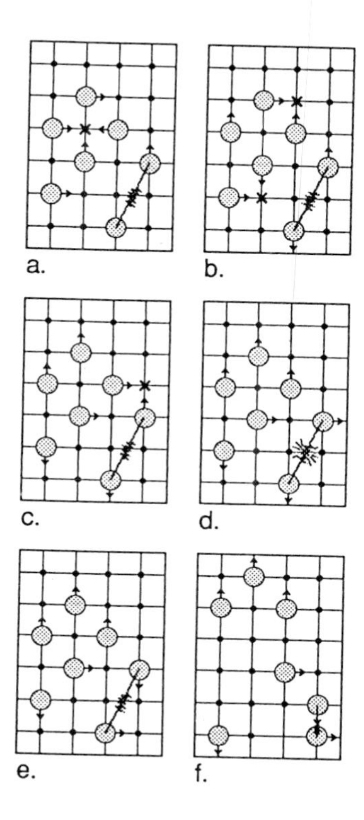

Fig. 1 Single time step shown in detail; a-d: Collision iteration (the crosses mark the sites that tend to become multiply occupied), d-e: chain-velocity update, e-f: propagation.

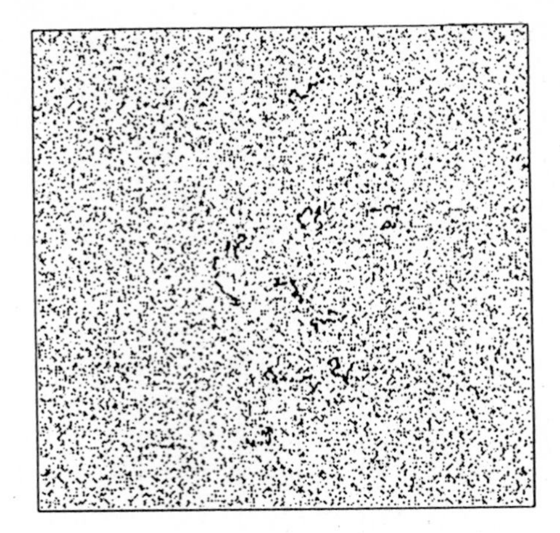

Fig. 2 Snapshot of a configuration during a simulation run on a 256x256 lattice in which 10 polymers each consisting of 40 particles (strings of black dots) move in a sea of solvent containing 15984 particles (grey dots).

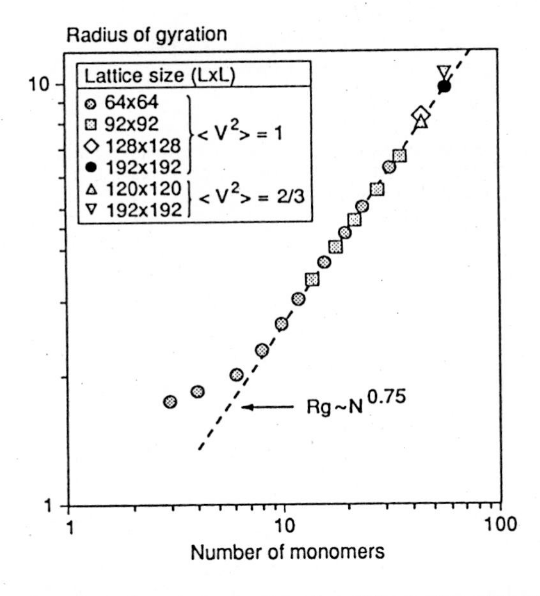

Fig. 3 Radius of gyration versus number of particles in the polymer chain obtained in 2-D simulations. The dashed line with slope 0.75 represents the theoretical scaling behaviour. The different symbols relate to different lattice sizes and different mean-square velocities of the particles as denoted in the legend. The particle density is fixed at a 25% site occupancy.

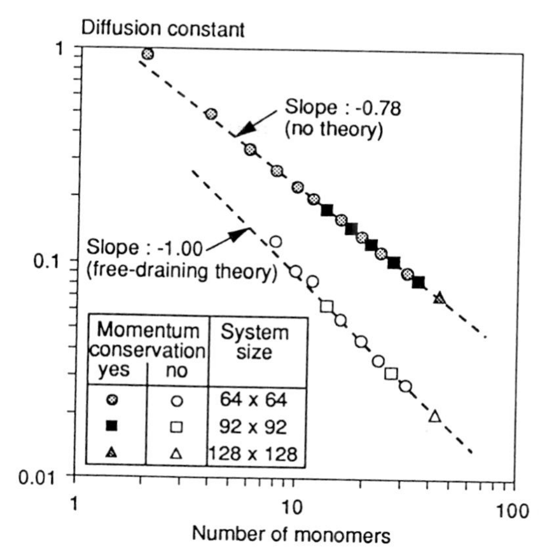

Fig. 4 Diffusion coefficient of polymers in a dilute 2-D solution as a function of the number of particles in the polymers. Closed and open symbols refer to collision rules obeying and violating momentum conservation, respectively. The particle density is fixed at a 25% site occupancy.

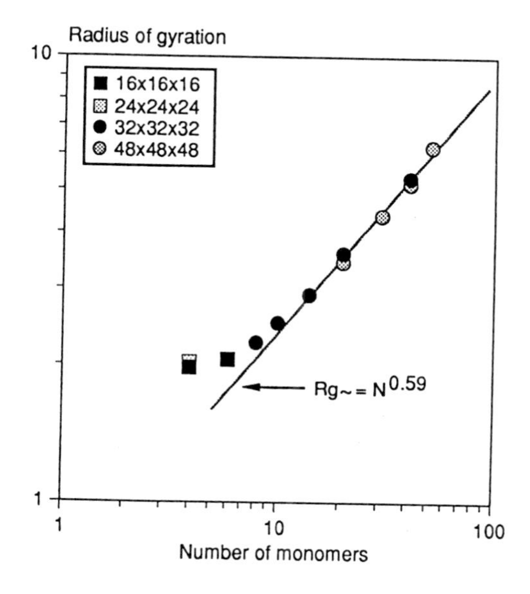

Fig. 5 Radius of gyration versus number of particles in the polymer chain obtained in 3-D simulations. The straight line with slope 0.59 represents the theoretical scaling behaviour. Different symbols relate to different lattice sizes, with the particle density fixed at 25% site occupancy.